Effective Environmental, Health, and Safety Management Using the Team Approach

Effective Environmental, Health, and Safety Management Using the Team Approach

Bill Taylor, CSP
Durham, North Carolina

WILEY-INTERSCIENCE

A JOHN WILEY & SONS, INC., PUBLICATION

For general information on our other products and services or for technical support,
please contact our Customer Care Department within the United States at
(800) 762-2974, outside the United States at (317) 572-3993 or fax (317) 572-4002.

Wiley publishes in a variety of print and electronic formats and by print-on-demand.
Some material included with standard print versions of this book may not be included
in e-books or in print-on-demand. If this book refers to media such as a CD or DVD
that is not included in the version you purchased, you may download this material at
http://booksupport.wiley.com. For more information about Wiley products,
visit www.wiley.com.

Library of Congress Cataloging-in-Publication Data:
Taylor, Bill
    Effective environmental, health, and safety management using the team
approach / Bill Taylor
        p.   cm.
    Includes index.
    ISBN-13 978-0-471-68231-8 (cloth)
    ISBN-10 0-471-68231-4 (cloth)
    1. Industrial hygiene.   2. Industrial safety.   3. Environmental health.
I. Title.
    RC967.T39 2005
    363.11—dc22

                                                            2005001261

10 9 8 7 6 5 4 3 2 1

Contents

Foreword

Knowledge of safety, health, and environmental issues and procedures alone will not protect people, property, and the environment. The key to success is the establishment of an effective safety, health, and environmental management system. Such a system, which involves all levels of the organization from the ranking manager to, and including, all employees, will continually increase awareness and protect people, property, and the environment.

Bill Taylor has had many years of experience in helping private and public sector organizations establish and maintain effective safety, health, and environmental management systems. This book describes Taylor's techniques, which organizations have used in establishing successful management systems.

The safety, health, and environmental management system described in Taylor's book is based on the same proven management techniques that organizations use to manage their other responsibilities, including production, service, quality, costs, and personnel relations. Those organizations successfully managing their other responsibilities can successfully manage their safety, health, and environmental responsibilities if they use the same management techniques. Taylor clearly describes how an organization can establish and maintain the necessary safety, health, and environmental management system.

Taylor's approach is no secret since it has been known for many years that when each level of the line organization effectively carries out its assigned responsibilities, the organization will succeed.

Thus, Taylor describes how all levels of the organization carry out their personal and organizational safety, health, and environmental responsibilities. Such a management system will effectively protect people, property, and the environment; comply with pertinent safety, health, and environmental standards, codes, and regulations; and protect individuals and the organization from liability.

The major emphasis of the system described by Taylor is top management commitment and participation. Since top management is the driving force to achieve all the organization's objectives, it must be the driving force to achieve the organization's safety, health, and environmental objectives. With the necessary management emphasis, the facility safety and health management system will succeed. Many top managers are not sure how to manage their safety, health, and environmental responsibilities. Bill Taylor describes an effective and user-friendly management system requiring minimum essential time that all managers can use successfully.

There are no organizational responsibilities more important than the protection of employees. Serious injuries and illnesses or death represent the ultimate organizational failure. Those organizations establishing and effectively maintaining the system described by Taylor will be successful and do what is reasonable to protect people, property, and the environment.

Raymond P. Boylston, CSP

Preface

For the first 12 years of my safety career, I had worked hard at managing employee safety and health programs with employers in private industry and local government. I achieved some minor level of success, won a few awards, and was generally satisfied with the way things were going. Actually, there was no cause for satisfaction as there were entirely too many injuries occurring, but as far as I knew, I was doing everything right as the safety and health manager. I figured that the problem was those stubborn employees who had their own agendas and the even more stubborn managers who viewed employee safety and health as a necessary evil rather than an integral component of the big picture.

The truth is that the managers did not know how to manage safety and health any differently. But then, neither did I. I was doing my job the only way I knew how and would later come to realize just how much better a job I could have done.

It wasn't until I went to work for Ray Boylston, Senior Vice President of ELB and Associates, that I began to learn how to manage employee safety and health.

In 1978, Ray Boylston, along with his two partners, Rick Ennis and John Lumsden, had formed ELB and Associates, one of the country's leading occupational safety and health consulting firms, and I felt truly privileged and very fortunate to be able to join the firm as a consultant and manager of training services. As a consultant with ELB I would have the opportunity to work under the direct tutelage of Ray, whom I would come to know as one of the greatest safety minds in the country.

In 1990, Ray published *Managing Safety and Health Programs*. By that time I had already been able to spend a great deal of time traveling with Ray to industrial sites and municipalities all over the United States and Canada and see first hand, how to manage employee safety and health. I was able to learn the same techniques that Ray had taught, not just as a consultant at hundreds of facilities across North America, but as a safety and health manager in a large facility that set a world record for going 13 years without incurring a single lost-time injury to any of its 2,700 employees.

It became clear to me, as I learned the system that Ray was teaching, that for 12 years I had done it all wrong. I became a disciple of the system and soon was turned loose by Ray to begin teaching the system myself.

For the first few years I was amazed at just how successful the system was in reducing injuries and illnesses to employees, but realized that the simplicity and logic behind the system made it a guaranteed success whenever it was implemented correctly.

It can truly be said that Ray Boylston wrote the book on occupational safety and health management. Now, it was time to update Ray's book.

I contacted Ray with the idea, and he was all for it, giving me carte blanche to do as I saw fit. When I contacted John Wiley & Sons, who had bought out Van Nostrand Reinhold, the original publisher, it was suggested that we do a rewrite instead of just updating the original. But as I began, I realized that there were many things in Ray's original that were so vital that they had to go into any subsequent publications. Besides, these were things I had learned from Ray and had been preaching myself for over 15 years. As they say, "If it ain't broke, don't fix it." So, while it is considered a new book on the subject, much of what Ray had stated in the original work is repeated here.

This book clearly explains this system and provides helpful information to anyone responsible for occupational safety and health, including security. It also provides insight for the reader into OSHA and OSHA regulations.

Well, Ray has now retired, although I still get to work with him on occasion, and ELB no longer exists. But I must say that I have had the great fortune to be able to continue to work with some of the country's most respected safety professionals. I am speaking of my two partners at CTJ Safety Associates, David Coble and Jim Jones.

I would like to acknowledge them both as individuals on whom I can always rely if I have a question about OSHA and OSHA law. I would particularly like to acknowledge David's contribution to this book. David created a set of very concise but extensive lists for Ray's original book and has updated them in this publication. If someone bought the book for the appendixes alone, it would be well worth the price.

Not everyone has a chance to learn from the best. I have made a career of it and am pleased to be able to pass some of that learning on to you, the reader. It is my wish that you will be able to take what is here and succeed in your efforts in injury prevention.

Bill Taylor, CSP

1

The Hierarchy of Safety, Health, and Environmental Management

When the plant manager sits down for the weekly staff meeting he/she is surrounded by those people with direct control and responsibility for all aspects of the business—department managers. If a problem comes up—say, for example, if costs are running too high—the manager does not turn to the finance manager, casting blame and dictateing that he/she work harder to cut costs. Instead, a good manager, knowing that cost overruns can occur anywhere, will try to pinpoint the problem so that efforts to resolve the problem can be properly focused. The manager knows that everyone sitting around that table has a responsibility to control costs and, likewise, is responsible for managing production or service, quality, people, and so on. These department managers in turn hold supervisors and employees accountable for the same charge. In other words, because any employee can damage product, affect quality, or run up costs, all employees must be held accountable for ensuring that production is met, quality is high, and costs are kept to a minimum. That is simply a part of the job;

Effective Environmental, Health, and Safety Management Using the Team Approach, by Bill Taylor
Copyright © 2005 John Wiley & Sons, Inc.

therefore, every employee makes up the production, service, quality, and cost organizations.

Why do we so often manage employee safety and health in different ways? At many facilities the safety–health manager is sometimes no more than a warm body. Often the ranking manager will assign responsibility for managing safety and health to an employee who is nearing retirement, or to a supervisor or middle manager who has little or no safety training as a collateral duty. Or perhaps the organization does go out and hire a bone fide safety professional; rarely is the individual a part of the manager's weekly staff meeting or agenda.

Even in large companies, who supposedly have good safety and health management systems, an individual who may be well qualified to manage will often be given the responsibility of managing environmental, health and safety (EHS) without the benefit of a staff, or authority to delegate. The individual's hands are tied and management effort is restricted to chairing the safety committee and making recommendations.

At one large facility employing over 2,000 workers, the plant manager and safety manager got into a heated disagreement. The safety manager ended up quitting that evening only to be replaced the very next morning by a young man who only the day before was a sewing machine mechanic in the same facility.

Did this plant manager consider safety so unimportant that anyone could take over and manage safety and health? And how much authority would he be willing to give to this new young safety manager? He would not be likely to seek advice from the young novice, nor would department managers or supervisors be likely to entertain his suggestions regarding issues within their respective areas of responsibility.

If it had been the human resources manager or the finance director who had quit, it's a sure bet that this young man would not have been appointed to either of these two jobs. The jobs are too important. They require a special knowledge and understanding of things beyond the mechanic's present education and background.

The message that the plant manager is conveying is that employee safety and well-being are of less importance than other functions. He does not view safety and health as something that should require education or special knowledge. But like many people, this plant manager does not understand the complexities of an effective safety and health management system, nor the specialized training and knowledge needed to effectively manage these systems.

This is a true story that actually did happen, and sadly it is more the norm when it comes to EHS management. We simply do not manage EHS with the same emphasis and enthusiasm as we do other functions. Even when it comes to enforcement, we fail to hold employees to the same level of accountability for using hearing protection, safety glasses, and lockout procedures as we do punctuality and insubordination. We do not enforce safety and health rules and procedures with the same emphasis as we do other rules and procedures.

At one facility the safety manager complained that there was a particular employee who would hitch a ride on the forks of the first passing forklift whenever he wanted to go anywhere within the plant. "I have told him every day not to do that yet he still does it and I don't know how to get him to stop," complained the safety–health manager. When asked what would happen if the employee were late for work, the safety manager stated that any tardy employee would receive a verbal warning on the first offense, a written warning on the second offense, suspension without pay on the third offense, and termination on the fourth offense.

So, why does this company not manage safety and health rules in the same way that they manage rules for punctuality? Does this mean that safety is less important than getting to work on time? It would be foolish to expect an employee to change his/her behavior, knowing that the convenience of the free ride meant more to this employee than did the consequences. In other words, the company did not enforce safety and health rules in the same way that they did other rules. They did not manage the EHS system the same way or with the same emphasis as other issues. Until they do, they can continue to expect safety violations to result in injuries and illnesses.

In every organization, regardless of how safe the tasks and facilities may be, employees can be injured or even killed while on the job. Since any employee can be injured or killed, every employee is responsible for safety. Thus, every employee makes up the safety–health–environmental organization.

THE SAFETY COP

A department head says to the safety manager, "You should have been here last night. John was working on the packaging machine and didn't have it locked out." The response of the safety manager is, "I can't be here 24 hours a day to police everything that goes on inside this plant." What makes the department head think that he did not have a responsibility to take action to enforce the violation and possibly prevent an amputation or fatality? The safety manager in this true-case scenario has become the safety cop.

The safety cop is the one who goes around the workplace day after day reminding employees to don safety glasses and earplugs; lockout equipment; or clear access to eyewashes, exits, and fire extinguishers. This should not be the routine of the safety–health manager. These are issues for which employees and supervisors must be held accountable. The employee must be held responsible for the condition of his/her immediate work environment, including any safety and health issues. They and they alone are responsible for ensuring that they are prepared to begin work each day by having proper protective equipment and that the equipment is properly locked out using personal locks and following all lockout procedures. It is part of the job and cannot be delegated to other workers. Any company that has a safety cop does not have an effective safety management system.

Safety–health managers should not be safety cops any more than the finance director or finance manager should be the finance cop going around monitoring the amount of waste, or if employees are damaging equipment or otherwise costing the company more money than is necessary. For one thing, this policing cannot be done effectively. That is the responsibility of the employee and the supervisor.

In order to make the workplace safe and meet OSHA compliance, there are literally hundreds, sometimes thousands, of machines, pieces of equipment, tools, processes, and other things that require inspection. Additionally, all employees require some level of safety training, records must be maintained and accidents investigated, noise and chemical exposures must be monitored, and written policies must be established. In other words, the job is too big for one individual or even a small team to accomplish. But even if an individual could do these things alone, this is not the way to manage safety and health. The safety cop style of management leads employees to believe that the safety–health manager is solely responsible for EHS. When workers do not have an active role in safety and health, they move away from their own EHS responsibilities and the EHS system.

While many managers will say that safety comes first, the reality is the bottom line, and bottom-line management often relegates safety to the back burner, in spite of all good intentions. And it could easily be argued that realistically speaking, safety cannot come before production or quality. After all, we don't build plants to have good safety systems. We build plants to produce a product or provide a service in order to make money. Without production or service, there won't be a need for safety and health since the facility will be closing down.

It could also be argued that it is more important to manage safety and health effectively than production, quality, and other management responsibilities, because the worst thing that could happen if we were to fail to effectively manage production would be for the company to go out of business or the plant to close. But if we failed to effectively manage safety and health, then people would loose their lives.

The question should not be which comes first or which is more important. Instead, we should manage all aspects of the business, including safety and health, equally. We should be asking if we involve employees in matters of safety and health to the same extent that we do production. Do we enforce safety rules and procedures with the same emphasis as other rules?

There will be times when a choice must be made between production and safety, but such choices will be made on a case-by-case basis as it is not always feasible to choose safety over production, especially in cases where the safety issue is deemed an acceptable risk versus a production issue that could bankrupt the company if not corrected. But the issue of managing safety and health is less about choosing between safety and production as it is ensuring that safety is managed as effectively. When we assign safety and health responsibilities to an individual and that individual is perceived as the safety cop, then we are not managing safety and health as we do other issues.

EMPLOYEE INVOLVEMENT

At one plant, the plant manager's executive secretary has prominently posted in the lobby, for all to see, a graph showing the rise in production that the plant has experienced over the first 6 months, since the new plant manager came on board. There is a great deal of pride in this accomplishment throughout the entire workforce because employees can see the numbers and take pride in their efforts. The plant manager is also proud because, after all, that is the very reason why he was hired, to improve production. And being the good manager he is, he has taken every opportunity to remind workers that this is the fruit of their hard work and it is their success as much as his.

During a EHS management system inspection an outside auditor pointed out a need for fall protection installed on the tops of equipment. The plant manager, concerned about the cost of installing rails, asked if there were an alternative and was told that until fall protection could be installed, workers who climb on top of the equipment would have to wear harnesses and lanyards with which to tie off. "Look," explained the manager, "our employees think that things like safety harnesses and belts are silly, and I can't make them wear things like that."

The auditor complimented the manager on his success in increasing production, asking how he accomplished such a feat. "We simply looked at where we were and laid out a plan to reach some goals, then told people what to do," replied the manager. "What would happen if the workers didn't do as you told them?" asked the auditor. "Well, we'd just get rid of them. I'm not going to lose my job like the last plant manager did." The auditor then pointed out how production and safety were not being managed in the same way:

> If you want to improve your safety you look at where you are, determine where you want to go and lay out your plan to get there. What you're telling me is that you're managing production one way and doing a great job, which was illustrated by the graph you have in the lobby. But you are managing safety in a different way and doing a poor job. That is illustrated by the number of employees you have walking around with eight or nine fingers, and the number of workers you have sitting at home right now on worker's compensation. If you will manage safety and health the same way you manage production, you will see a similar improvement.

Not long afterward the auditor was back in that same plant installing the system described in this book. Less than 6 months later, injuries were down by 50% at this location.

Getting employees involved in the safety–health process is paramount to the success of worker safety and health management systems. Otherwise, workers view safety and health as someone else's responsibility and view the safety manager as the safety cop. When we get workers involved in safety and health activities, we increase their awareness. It is this increased awareness that leads to fewer injuries and illnesses in the workplace. This is how we build a safety culture, and until we have this culture, employees, supervisors, and managers will continue to violate rules pertaining to safety and health as they view safety and health as someone else's responsibility.

Employee involvement can be achieved by issuing safety and health assignments and giving employees and supervisors something to do to get actively involved in the EHS effort. There are several advantages in doing this, which will be discussed in detail in the chapters that follow.

Suffice it to say that when we achieve employee involvement and increased participation, we will reduce workplace injuries and illnesses, get closer to achieving OSHA compliance, reduce costs, and establish a safety culture.

After all, someone's safety is a personal matter, and it is the personal responsibility of the worker to do the job right, including doing the job safely. Many times managers view the workforce more as an institution than as a group of individuals. Employees are not buildings or machines. The workforce is not a book club with unidentifiable members scattered around the country whose only existence is by way of a number in a computer; they are humans with feelings and needs and families. It is very easy to lose sight of this fact and inadvertently become callus to the needs of the workforce, including their needs for protection on the job. But workplace safety and preventing injuries is a personal matter; thus it is important to view safety and health on a personal level.

A manager at a small plant of ~200 employees who was accompanying a safety consultant on a plant safety inspection was told that he could cut employee injuries in half within 4–6 months using the "facility safety and health management system." His immediate response was "What's the cost?"; not "Tell me more" or "That sounds interesting" but "What's the cost?" He was given a quote and quickly balked, saying that he just couldn't justify that kind of expense. He proudly pointed out that the facility had only five recordable injuries the year before and only one of them was serious—an amputated finger, at which point the consultant asked the manager to look around at his employees on the floor busily going about their duties, and pick out the five he wanted to get hurt this year.

The perspective changes somewhat when put on a personal level. Naturally, the plant manager wouldn't want a single employee to have an injury, especially an amputated finger. But this manager was willing to accept injuries as a part of doing business rather than to make a small investment to put a stop to them. Would he be any more concerned if his own finger or one of his children's fingers were amputated?

Even in those facilities where the plant manager can and does go onto the plant floor and greet workers by name on a daily basis, if management does not recognize the need to manage safety and health like other matters, then injuries are regretful, but still accepted as a part of doing business.

Injuries and illnesses should not be accepted as a part of doing business any more than production shortfalls or budget over-runs are accepted. Such things are not acceptable and will not be tolerated.

2

Understanding OSHA and Safety and Health Regulations

In 1970 former President Nixon signed into law the Occupational Safety and Health Act. Now, for the first time there was a law that would require employers to provide safe and healthful environments for every working man and woman in the nation. Many employers were already providing safe workplaces, having long since recognized the moral and ethical obligation to protect workers. But this could not be said of most employers. In fact, prior to OSHA, safety programs ran the gamut from nothing at all to the well-established and elaborate programs of the Du Ponts, the ExxonMobils, and the Alcoas.

While many employers viewed OSHA with fear and loathing, those who were practicing good safety and health were prepared to have them come in and in many cases would actually welcome the OSHA inspector.

The OSHAct (OSHA act) did several things. In addition to establishing the Occupational Safety and Health Administration (OSHA), it gave the administration the authority to not only make law but also incorporate consensus standards by reference. This was very important. It meant that many of those voluntary standards from the American National Standards Institute (ANSI), the National Fire Protection Association (NFPA), and other organizations would now become law. Overnight, with the stroke of a pen, hundreds of consensus standards went from being voluntary to mandatory. For this reason, if an employer seeks OSHA compliance, it is not enough just to rely on that copy of the OSHA General Industry Standards that has been collecting dust on the shelf since 1970. Employers must also become familiar with the consensus standards applicable to their operation. You can learn more abut these consensus standards and how to access them by going to the Websites identified in Appendix C.

The first thing one must learn about OSHA standards, including those incorporated by reference, is that they are the bare minima. Some of these standards were established as much as 100 years ago, and while they may have been adequate at the time, they are no longer as effective as they once were. A good example is the current OSHA standard on machine guarding, which tells us that equipment that is more than 7 ft high is considered to be guarded by location. At one point in time 7 ft may have been high enough to prevent employees from reaching the hazard; however, in the twenty-first century we are taller and almost everyone can reach to a height of 7 ft. This means that although we may be in compliance by employing the 7 ft rule, we are not as safe as we could be. Thus, it is better to concentrate on employee protection as opposed to OSHA compliance. This may mean exceeding OSHA standards in many instances, but it will afford better protection for workers.

OSHA has codified standards into six categories. There are OSHA standards for agricultural employment (1928), ship building and ship breaking (1915), marine terminals and longshoring ashore (1917), longshoring afloat and maritime (1918), construction (1926), and general industry (1910). The majority of employers in this country are general industry employers and thus must comply with the general industry standards found at 1910. That is not to say that general industry employers need not be concerned with other standards. For example, at 1910.12, construction work is defined as "work for construction, alteration, and/or repair, including painting and decorating." Given that, there will be times when employees of a general industry employer will perform work that is considered as construction, thereby requiring them to comply with the construction standards found at 1926, in addition to the general industry standards. An example would be when the maintenance department at a furniture manufacturing plant, a general industry employer, builds a new office for the maintenance manager. This work would fall into the realm of construction requiring compliance with construction standards, meaning that the employees would be required to use ground-fault circuit interruption protection for extension cords and portable electric equipment, or have an assured grounding program. Such protection is not required under the general industry standards. Therefore, as they go about building the new office, they will be required to comply with both construction and general industry standards.

It is sometimes a very fine and, at times, indistinguishable line between what is construction and what is general industry—still more reason for exceeding OSHA requirements.

The OSHA standards are intended as a guide for employers to use in order to provide a safe work environment. The standards spell out instruction on how to properly guard machinery, work safely inside confined spaces, operate forklifts safely, and so on. There are literally thousands of standards to which employers must adhere. Still, standards do not cover every conceivable situation. With over six million employers in hundreds of different industries, this would not be feasible. For this reason, the OSHAct contains, at Section 5, paragraph (a)(1), what is referred to as the "general duty clause."

The general duty clause is a tool to be used by OSHA to require that safety precautions be taken to protect workers from hazards that are likely to result in serious injury or illness in the absence of a specific standard. For example, an OSHA inspector was walking through one facility when he observed a bird flying around inside the building where there were employees working. Because of the potential health hazard, the employer was cited using the general duty clause since there is no standard prohibiting the presence of birds inside a facility.

Ergonomics is another example of how OSHA uses the general duty clause. The presence of ergonomic stressors in an employee's daily tasks can sometimes lead to such disorders as tendonitis or carpal tunnel syndrome. These are serious conditions that can often be prevented by training or making changes in the task or the workstation. In the absence of an ergonomics standard, failure on the part of an employer to recognize and deal with such stressors has been cited by OSHA using the general duty clause.

In 1990, in an effort to help offset the budget deficit, OSHA was instructed by Congress to collect more money in penalties, and in essence, to be more aggressive. As a result, we saw OSHA jurisdiction extended to more public sector employers such as cities, counties, public universities, and the U.S. Postal Service. We also saw the use of standards that prior to that time got little use, as well as the creative use of many other standards.

For years OSHA did not require employers to label storage racks to indicate maximum storage capacity. Now, however, OSHA issues citations where this has not been done. The standard used to cite this, 1910.145(c)(3), is an old standard that requires the use of safety signs to warn of hazards. The standard says nothing about storage racks or storage capacity. If employers are not aware of such issues, then they are at risk of being cited by OSHA. But worse still, they are not likely to provide the warning needed to protect employees.

Because there are over six million employers in this country involved in hundreds of different types of industry, writing safety and health standards becomes a daunting task. It would not be practical to write a standard, such as the confined-space standard, and apply it to all employers in all types of industry with the same expectations for every confined-space entry. Look at atmospheric testing, for example. It would not be reasonable to include in the standard a requirement that employers test the atmosphere in a permit confined space at least once per hour. In many spaces this would be overkill and an expensive waste of time. In some cases, however, hourly testing would not be sufficient and would put workers at risk of injury or death, unless continuous testing were conducted. For this reason there are two types of standards: specification standards and performance-based standards.

A specification standard is one where there is no subjectivity. Everything one needs is clearly stated in black and white. The standard on fixed ladders is a good example. At 29 CFR 1910.27(b)(1)(i)–(iii), the standard gives specific information regarding the diameter and length of and spacing between rungs on fixed ladders. This is not subjective, nor open to debate or discussion. The standard states that ladder rungs shall not be more than 12 in. apart. This means that if the rungs are 13 in. apart, the employer is in violation.

Performance-based standards, on the other hand, leave much of the decisionmaking to the employer. Go back to confined-space atmospheric testing. Rather than try to tell employers how often to test the air under varying conditions, OSHA simply leaves it to the employer to decide, based on the hazard and hazard potential, how often is often enough.

Performance-based standards, however, create yet another problem: interpretation. If employers are left with the decision-making authority, there must be a way for them to evaluate their decisions to ensure that they are making the right ones—not just for safety sake but also to ensure that they are in compliance. It would be very easy to interpret a standard one way only to have an OSHA compliance officer come along and cite the employer because he/she has a different interpretation. For this reason OSHA provides directives.

Two types of directives are issued by OSHA. The STD or standards directive is intended to help employers interpret standards. The STD also is used to inform employers about changes, exceptions in the standards, and so forth. The other type of directive, the CPL, or compliance directive, offers interpretation as well, but its primary intent is to assist OSHA compliance officers in enforcing standards uniformly. By adhering to CPLs, the compliance officer in Texas should be interpreting and enforcing the standards in the same way as the compliance officer in Georgia does.

Another tool used by OSHA is the letter of interpretation. When an individual writes a letter to OSHA asking a question, OSHA responds with the answer in a letter that states their position on the issue in question. OSHA has issued thousands of letters of interpretation, which, along with the directives, can be accessed at the OSHA Website.

The point of all this is that there are literally thousands of safety and health standards that employers must recognize and adhere to. Without knowledge of where to look, finding the standards can be difficult enough; understanding them and how they may apply to a given situation can be much more difficult. The information is available to anyone. In this day of computers and Internet technology, it is usually simply a matter of asking. But because of the volume of information that is available and required, there should be someone on staff responsible for staying abreast of what is important. But it quickly becomes obvious, too, that the job of meeting all the OSHA requirements, even in a small facility, can be overwhelming. If the goal is to manage safety and health in the same way that we do everything else, then there will be others to help with the overwhelming burden of implementing this information, and this individual becomes more of a coordinator, providing leadership in safety and health.

Just as the employer will employ the services of an expert in tax law or finance, the employer should also have someone on staff who is familiar with workplace safety and health law, or at least who knows whom to contact for assistance.

The FSHC (Facility Safety and Health Committee) managing system described in the following chapters is intended to get employees at every level within the organization involved in the safety–health program.

3

The Basics of Managing Safety, Health, and Environmental Programs

Employers establish teams, such as quality assurance (or control) teams, to get employees involved in the quality process. We empower employees with the freedom to stop an entire production line if they become aware of a problem affecting production or quality—and because of this increased participation our production is up, quality is good, and costs are down. We have sought the participation of a very vital and knowledgeable resource—our employees. But again, we manage safety differently. We do not involve employees in the safety process except maybe to put them on a safety committee. We fail to give them responsibilities in an effort to get them involved in the safety–health process.

But we have learned that if we want to improve something for which employees are responsible, then we make it an important part of their workday by making it a part of their job. We put them on a team or committee and give them responsibilities. We hold them accountable and recognize their efforts. By involving them, employees develop a greater sense of awareness and ownership.

Effective Environmental, Health, and Safety Management Using the Team Approach, by Bill Taylor
Copyright © 2005 John Wiley & Sons, Inc.

Results include better production and lower cost. This empowerment (involvement) is actually changing the behavior of the workers. We should, likewise, strive in the same way to change unacceptable safety behaviors. Similarly, we must hold managers and supervisors to the same level of accountability. It does little good to punish a worker who violated a safety rule and at the same time turn a blind eye to the supervisor or manager who sanctioned the violation through his/her silence. The job of the supervisor or manager is to ensure that the job is done right. If the job is not done safely, then it is not done right. And if the manager or supervisor is aware of this and fails to respond, then he/she has not done the job right, either.

Managers are nearly always a part of the safety problem; therefore, they must also be part of the solution. There cannot be long-lasting safety success unless the management team is a part of the safety–health effort. The goal of every organization should be to build a safety culture, and this can be accomplished only through employee involvement.

By getting employees involved in doing inspections, investigations, and other procedures, the needs of the safety and health program can be met and employee safety maximized by building a safety culture through the increased awareness on the part of employees.

MAXIMIZING SUCCESS

To maximize success in business, whether in manufacturing or the service industry, we must recognize and effectively manage all components of the business. Some of those components include production (or service), quality, costs, personnel relations, and safety and health. It does little good to manage only some of these key issues well, and simply do a mediocre job managing the others. Oh, there may be good production, or quality may be exemplary, but unless all matters are managed in the same way, with the same emphasis, the overall success of the organization will be compromised.

A good manager will make every effort to address and manage the business in its entirety, although it is not unusual for a manager to excel in certain areas. Case in point is the plant manager mentioned at the outset who had established a reputation for improving production wherever he went. While this manager had proved to be a strong champion of production, exhibiting strong management skills, he was lacking when it came to managing safety, health, and environmental matters. This was evident in the number of employees who were injured at his facility. This is not to say that he was incapable of managing safety well; it's just how he had things organized and his narrow focus on production. All he needed to do was recognize safety, health, and environmental programs as being vital to the overall success of his organization and then manage these things in the same successful way that he was managing production.

Managing anything successfully mandates adherence to certain elements. Employers must

- Plan
- Organize
- Lead
- Control

Regardless of the function, all management systems must have planning, organization, leadership, and control in order to succeed.

In the workplace, there are numerous things that must be successfully managed for the company to succeed. And the better we are at managing these issues, the greater the company's success. Every employer is concerned, and rightfully so, with production. Or, if the company is a service provider, then service is vital. Likewise, employers are concerned about the quality of production or the service they are providing, for without good quality, the company is not likely to succeed.

Cost is equally important, and employers work to find every advantage and method to reduce cost. Without the workforce, none of this would matter, anyway. If we have employees, then we must effectively manage people. Finally, we must focus attention on safety and health of the workforce. Employers can neglect safety and health, but when they do, injuries and illnesses are the result and this affects the bottom line as it sends insurance and medical costs up, we suffer manpower shortages, and worst of all, the very life of the worker is jeopardized.

So, how do we manage these issues? We begin by planning. Like taking a trip, before you begin, you plan the trip. You need to lay out the course, beginning with where you are, your destination, and then how you plan to get there. Which roads will you take? Where will you stop along the way? How long will it take to reach your destination? These are important questions when planning the trip, but they are equally important in planning production.

It is important to identify the present location. Where is production in terms of where we want it to be? Are we currently able to produce enough product to meet the demands of our customers? If we want it to be better than it is, then we must determine where we want to take it. In other words, what will be the destination? What production goals have we established? Knowing our goals will allow us to determine what must be done to reach them.

We must then have an organization to carry out what it is we are planning. We need people to run machines, drive forklifts, clean and maintain machinery, supervise, and perform other tasks. Regardless of what their jobs might be, everyone within the workforce has a responsibility toward production. Therefore, every employee makes up the production organization.

To help keep us on track, we assign a leadership role to someone we may call a "production manager," a "vice president of production," or any of a host of other names. But this is the person who provides production leadership and lets the workforce know what the organization's expectations are in meeting production demands so that we can achieve our goals. This is also the person to whom we go when there are production issues that need a management decision.

We also must maintain some sort of control over production. We may establish quotas or reward individual accomplishments to recognize outstanding production effort on the part of an individual employee or a group of employees. We have to exhibit control to keep within our production plan; otherwise there may be a loss in revenue as costs go up.

Equally as important as production is quality, for without good quality, production will be less important since we will be losing customers. To maximize quality, we do some quality planning. We determine exactly what it is in terms of quality that our customers expect and then fine-tune the quality of the product and produce according to that plan.

Since any employee can affect quality of our product or service, all employees make up the quality organization. There is usually a quality control manager who assumes leadership for the quality effort. This individual manages the quality assurance group who provides guidance and assistance to ensure that we stay within the quality guidelines established in the plan.

Quality is closely scrutinized to ensure that we are producing at the best possible level of quality. To help identify when there might be quality problems, we take random samples of the finished product and test them extensively, again, to keep us within the predetermined parameters. This is quality control.

No business will last very long if costs get out of control. To help us manage costs, we begin in the same manner. We plan our budget for the coming year. We have to determine how much money we will need to operate next year. This is cost planning and is vital for success.

Who can cost the company money? Anyone can damage product. Anyone can damage equipment or make a customer angry, thus affecting the company's reputation. In other words, anyone can cost the company money, and all employees have a responsibility to stay within our cost plan, or budget. Therefore, everyone makes up the cost organization.

However, our employees are not cost containment or cost management experts. Cost leadership comes from our controller, finance director, or whomever is charged with the responsibility for overseeing the budget. It is this person to whom the rest of the company will turn for direction on spending and costs.

Like everything else, we must manage our workforce. Most new employees at every company will be given an employee's handbook or similar publication that clearly spells out employee benefits, insurance, vacation and sick leave policies, and so on. At some point someone within the company had to sit down and plan what these policies would say and develop the handbook. This is personnel planning.

Now we are talking about people. People must interact on a daily basis. They should work as a team to achieve the goals of the company, which is maximum production with highest quality at a minimum cost. So, it goes without saying that everyone makes up the human resources organization.

Leadership for personnel matters comes from the human resources department. It is this group of employees, usually led by a human resources manager, who is well versed in the intricate laws as they relate to hiring and terminating employees as well as the other policies regarding employment.

We control employees when we tell them what time to come to work, what time to go home, how long to take for lunch, and so forth. We even go so far as to define the immediate family so if an employee's great-aunt Bessie dies, the employee can look up the definition of "immediate family" in the funeral leave policy to determine whether he/she must take vacation to attend Bessie's funeral.

As for safety and health, if we plan to manage them for maximum success, then we have to manage the program in the same way. We must plan our safety efforts by determining where we are and where we want to go. Are our employees getting sick or injured on the job?

A safety–health plan will include what committees will be established. What will people, employees, and managers do within the safety effort? What meetings will be held? Who will conduct investigations or training? An effective plan will be our roadmap to take us where we want to go with safety and health.

Since any employee can be injured or even killed in the workplace, and since all employees are responsible for safety, every employee makes up the safety–health organization. This organization does not consist solely of the safety manager, industrial hygienist, and/or plant nurse any more than is the production manager the sole employee responsible for production.

Instead, safety–health professionals provide the leadership in safety and health. They are the experts to whom we turn when we have questions related to safety and health. They coordinate efforts and keep top management informed on matters related to safety and health.

Safety–health control is provided in several ways. Policies and procedures, along with signs and warnings, provide some measure of control. The level of control is only as effective as the level of enforcement of the policies. Where enforcement is weak, control and thus compliance are weak as well.

Signs are used as a means of controlling the speed limit on our nation's highways. But only where the signs are strictly enforced do we see drivers complying with the indicated speed limits. In most instances drivers will drive as fast as they think law enforcement will allow. Thus, it is not the sign that controls speed on the highway; instead, it is a level of enforcement established by local law. It is how much an enforcement officer allows drivers to exceed posted speed limits. So, the sign offers very little in the way of control.

The same people who exceed posted speed limits also work in facilities where rules may be indicated by the use of signs or written policies, for instance, "Hearing protection required in this area," or "Lockout machine before opening guard." And those people comply with the signs or rather fail to comply with the signs in the workplace in the same way they do when out on the highway, and for the same reason; because, as on the road, there is also a level of enforcement inside the workplace. That level is determined by management.

Does the supervisor strictly enforce the proper use of hearing protection? Do supervisors strictly enforce lockout, or do they allow employees to forgo lockout procedures as a means of saving time? Employees will comply with rules to the extent that rules are enforced. If enforcement is weak, then compliance will be weak.

In order to ensure maximum production with top quality and good cost control, we must maximize management efforts of these components. Likewise, to maximize safety compliance to prevent employee injury and illness, we must maximize management of safety and health. To do otherwise is to say that employee safety is not important, or at least less important than other issues.

THE FACILITY SAFETY AND HEALTH COMMITTEE

Good managers encourage strong communication within the organization. They know that each department must know what the other departments are doing. They know, too, that the organization must be aware of management expectations in order to carry out the desired plans and activities. For this reason most facility managers begin the week by meeting with the management staff. This usually consists of department heads and other key personnel. It is during these weekly staff meetings that policy is established, problems are discussed and resolved, and the manager sets the tone for how the operation will function. This weekly meeting is not a planning meeting so much as it is an informational meeting that allows the manager to keep his/her hand on the pulse of the organization.

The department managers, in turn, have been authorized to return to their respective departments with the information garnered during the meeting and implement the manager's instructions.

If our desire is to manage safety and health in the same way that we manage everything else, then it follows that there should be some gathering of department managers reporting on issues of safety and health to the plant manager. For this reason, the Facility Safety and Health Committee is made up of department managers and is chaired by the ranking manager.

By structuring in this way, several positive things are accomplished:

1. As department heads report to the committee on matters of safety and health, it gives the plant manager the information needed to stay abreast of what is happening and make proper decisions in this field.
2. Instruction and delegation of safety efforts is coming from the plant manager and not the safety manager. This is very important since few safety managers have authority over department heads.
3. Finally, with the plant manager chairing this committee and making safety and health a routine part of his/her agenda, a message is being conveyed to managers, supervisors, and employees that safety and health is an important issue. If an issue is important to the ranking manager, then it is more likely to be important to everyone else within the organization. But if the organization perceives an issue is of less importance to the manager, that level of importance is reflected in how the issue is managed throughout the organization.

Therefore, FSHC system effectiveness is dependent on the effectiveness of management of the organization and who chairs the FSHC. The greater the authority of the FSHC chairperson, the more effective will be the FSHC system. There is little doubt this committee should be chaired by the ranking manager.

4

The Facility Safety and Health Committee System

PURPOSE AND ORGANIZATION

The FSHC system is made up of nine committees known as *task groups* and a tenth committee called the *Facility Safety and Health Committee*. The purpose of the groups that make up the system is to manage the organization's overall safety and health system.

It is within these task groups that the work of pursuing employee safety and OSHA compliance through effective management takes place. Rather than expect the safety and health manager to go around the facility conducting the thousands of inspections required by law, enforcing the use of PPE and lockout procedures, conducting investigations, leading safety meetings, conducting training, and doing everything else needed to effect employee safety, these responsibilities are turned over to employees and supervisors sitting on the task groups. The organizational chart in figure 4-1 illustrates how the task groups fit into the system.

Each task group is made up of employees and supervisors and chaired by a department manager. This allows for a significant

Effective Environmental, Health, and Safety Management Using the Team Approach, by Bill Taylor
Copyright © 2005 John Wiley & Sons, Inc.

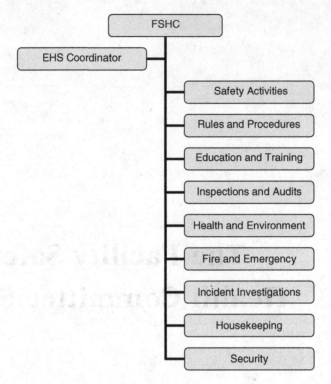

Figure 4-1. Organizational chart.

number of employees to become directly involved with safety and health. This is how the participation level, and thus the awareness level, is increased. This is what builds a safety culture.

GETTING STARTED
SELECTION OF PARTICIPANTS

Contrary to common belief, the enrollment of a significant number of volunteers to fill task group positions is rarely a problem. Experience has shown that employees are more than willing to get involved in the operation of the safety–health process and are usually quick to volunteer. However, as selection is made, consideration should be given to the interests and expertise of potential participants. Employees should be permitted to volunteer for participation on the task group that interests them but should be apprised of the mission of each task group and what might be expected of individual members. This will help to match members with their areas of interest.

A major concern, especially at the outset, should be the selection of task group leaders who exhibit strong management and organizational skills. It is vitally important to get the system off to a good start, and this becomes more difficult if task group leaders have weak management skills. Weak leadership can lead to apathy on the part of task group membership, which can lead to dysfunction and eventual collapse of the task group or the system. As task groups become firmly established, it becomes less of a threat later when department heads having less than strong management skills are assigned responsibility to lead a task group.

The skills matching chart in figure 4-2 is intended to help the employer match employee skills with task groups. It should be helpful for task group leadership.

TRAINING PARTICIPANTS

The facility manager and department heads should first be given an overview of the system. The overview should be a 35–45-minute synopsis of the need for an effective management system and how the FSHC system functions. The purpose of the overview is to

- Provide information enabling the organization to assemble task groups
- Display the ranking manager's commitment level to department heads
- Clear up misconceptions regarding the system and lay the groundwork for the training of participants

Following the management overview, all supervisors, department heads, and the facility manager receive a 3-hour training session covering the following:

- Current trends in safety and health
- Managing with the FSHC system
- Hazard recognition

The first part of the training is intended to make trainees aware of growing public and individual concerns regarding safety, health, and environmental issues. The second session describes the FSHC system and how it is intended to function. Finally, the last session covers hazard recognition using pictures usually taken within the

Task Group	Criteria	Comments
Safety Activities	Ability to create, sell, promote, motivate. High initiative with many good ideas and suggestions.	Must be able to analyze and evaluate the effectiveness of the entire safety and health program.
Rules and Procedures	Good knowledge of safety rules. Strong advocate for enforcing rules and procedures. Aware of the need for rules and procedures.	Must coordinate the annual review of all safety rules and procedures.
Education and Training	Strong believer in the benefits of education and training. Someone other than the staff training manager/coordinator. Effective in setting priorities.	Must have strong coordination skills.
Health and Environment	Good knowledge of chemicals and chemical hazards. An awareness of occupational diseases/illnesses such as ergonomic-related illnesses. A strong environmental awareness.	Should have a team member with strong ergonomic knowledge to chair an ergonomics team. Also, a member to chair an environmental team.
Inspections and Audits	Strong awareness of the need for and the benefits of routine inspections. Good safety and health knowledge. Knowledge of standards, codes and regulations.	Must be able to report on the inspections performed by others and to identify repeat defects.
Fire and Emergency	Strong engineering, maintenance background. Usually the engineering or ranking maintenance manager. Knowledge of fire prevention, control and emergency response, including hazardous materials response.	Coordinates all fire brigade and emergency team activities. Needs knowledge of OSHA emergency standards.
Accident Investigation	Strong analytical ability. A stickler for details. Knows the importance of obtaining the facts. Understands that practically all accidents can be prevented. Knows that effective management systems can prevent most accidents.	Must be able to report results and facts accurately and effectively.

Figure 4-2. Skills matching chart. *(continued)*

Housekeeping	A believer in the need for, and importance of good housekeeping. Understands the role of management leadership in housekeeping.	Reports housekeeping audit results accurately. Comments on good housekeeping, points out problems, and offers constructive suggestions for improvement.
Security	Understands the importance of protecting people and property from the threat of violent acts. Realizes the potential for sabotage and espionage. Good organizational skills.	Maintains alert status and reports potential threats. Maintains rapport with local authorities. Reports security audit findings and recommendations.

Figure 4-2. *(continued)*

organization's facility. This gives participants the opportunity to see hazards in a different light and gain an understanding of what constitutes a hazard. Best of all, the final session is effective in increasing each participant's level of safety awareness.

This training is followed by a 90-minute training session with each of the nine individual task groups and their chairperson. This training covers the purpose of the task group and how each one goes about accomplishing its mission, including individual member responsibilities.

MANAGING THE SYSTEM
MEETINGS

Each task group should meet once per month. The meetings should last only about 30–45 minutes. The purpose of the meeting is so that one member can give a brief report on the status of his/her assignment. It is vitally important that meetings never be canceled or postponed. When this begins to happen, it doesn't take long before the group begins to fall apart and no longer conducts meetings.

It is suggested that meetings be scheduled on the same day of the month at the same time and at the same location. By scheduling meetings in a routine fashion like this, it becomes much easier for members to remember meetings and to schedule other events around them. A form is available at Figure A-1 (in Appendix A at the end of this book) to assist task group leaders in meeting scheduling.

In the event a member cannot be present, even if it is the chairperson, the meeting should proceed as scheduled. If the chairperson cannot be present for some reason, then a member of the task group should fill in and chair the meeting that month. If the member who is scheduled to report that month cannot be present for some reason, the meeting goes on as scheduled but another member should give the report in his/her place.

AGENDA

Every meeting should follow an agenda. An agenda helps keep the meeting on track, adhering to the subject. Meeting topics should be determined at the beginning of the year so that each member knows beforehand what will be discussed and when. Additionally, the chairperson should send out an agenda to each member prior to the meeting as a reminder.

Go back to the inspections and audits task group member mentioned earlier. Her assignment is forklifts. She has completed her audit of forklift inspections for the month and discovered that needed repairs are not being made in a timely manner. Her task group is scheduled to meet this week, and the topic of discussion is scheduled to be ladder inspections. Her topic, forklifts, is not scheduled until 6 months later. Rather than wait for 6 months, she would mention her findings in the next meeting so that this can be brought to the attention of the FSHC in their next meeting, and to the appropriate manager responsible for forklift repair immediately. Then, when forklift trucks come up as the monthly topic 6 months later, she can include this information in her report along with anything else that has come to her attention since that time.

It would be the responsibility of the task group chairperson and not the employee to convey this finding to the manager responsible for forklift truck repair. Likewise, the chairperson would report this finding to the FSHC at the next scheduled meeting.

MINUTES

Every meeting should be documented, with someone keeping minutes to provide a written record of who was present and what transpired. A blank form that can be used for this purpose is shown in Figure A-2.

All members should receive copies of their task group minutes. Also, the FSHC should get a copy of minutes from each task group. Likewise, each task group should get a copy of FSHC minutes.

It is also suggested that prior to each meeting of the FSHC, the plant EHS manager meet with the plant manager (FSHC chair) and go over the minutes from each task group. This will bring the manager up on where each group stands and is a good way to prepare the facility manager for the FSHC meeting. Knowing that the minutes will be reviewed by the plant manger also helps to ensure that the minutes will be accurate.

ASSIGNMENTS

Each task group member is given an assignment by the task group chairperson. The assignments are among the most critical components of the system. By issuing assignments to task group members, several things are accomplished. First, as the employees get more involved with the assignments, their knowledge will increase. Employees will continue to work with the assignments, thus becoming the resident experts in their given assignments. This means that if there are five members on each task group, there are 45 employees actively involved in safety and health, each becoming proficient in their assigned areas.

With this increased knowledge there is now someone other than the EHS manager capable of reviewing or performing activities required by OSHA. The daily activities that go into meeting OSHA requirements and protecting employees is now distributed among scores of employees, each spending a minimal amount of time each month to ensure that these requirements are met.

The real value in issuing assignments comes in the increased awareness of the employees on the task groups. By working on assignments, employees not only become more aware of that particular issue but also develop a greater safety awareness overall and thus become safer workers.

Most people have experienced increased awareness regarding a particular issue as a result of their participation in something. How many times has the safety professional gone into a restaurant and pointed out to his/her spouse the table blocking an exit or fire extinguisher? The spouse didn't notice that. The one who invests the time each day learning standards and managing safety was instinctively aware of it, however.

Or, consider prospective car buyers. As they begin to shop around for a new car, they visit several dealerships over a period of several days. They test-drive several cars and invest considerable time in the tedious process of buying a car. Finally, they make a decision and sign all the papers. Then, on the way home in that new car they begin to notice all the cars on the road just like the one they just bought.

Those other drivers didn't just go out and buy their cars at the same time. They already had theirs, and had them on the road. They were there all along. But the awareness has now increased as a result of the time invested and the knowledge gained during the buying process. Through no effort on the part of this participant in the car buying process, he/she is now much more aware of this type of car. All this car buyer had to do was to get involved.

As employees begin to get involved performing their assignments, they will develop a similar increased awareness and will be safer workers. They don't have to become safety and health experts. They don't have to take a test. All they have to do is get involved.

Two things should be mentioned regarding these assignments. The first thing is that they are permanent for as long as the employee serves on the task group. The assignment is not something that the employee will work on and complete after a month or two then come back for a new assignment. An employee's assignment, for example, might be to audit ladder inspections. Well, as long as the employer has ladders, there will always be a need to conduct ladder inspections. And as long as there are ladder inspections, there should be someone to audit to ensure that they are being inspected properly.

The second thing is that employees should not devote too much time to assignments. This is a common concern, especially among supervisors. It is understood that everyone has a full-time job already. They cannot afford to take an hour each week and devote to their assignments. With 45 or 50 people involved in safety and health, it should not be necessary to spend a great deal of time on such matters. Instead, task group members should spend only whatever time is reasonable and affordable on assignments. The time will vary depending on the assignment, but none should require more than an hour each month. Most will require far less time than that.

For example, a member of the Inspections and Audits Task Group is assigned the topic of forklifts. That means her role is to audit the inspections the fork truck drivers perform each day. She does not physically go out and do an inspection. Instead, she might collect all the daily forklift truck inspection checksheets that each driver has turned in during the previous month. She would then spend a few minutes looking through them to determine the quality of inspections being conducted. She may see where the same inspection finding was written up each day by the inspecting driver. This would indicate that the forklifts, or at least that fork-lift truck, are not being taken out of service so that repairs can be made.

Or, she might notice that instead of checking each individual box on the checklist, the driver made one check in the top box and drew a line downward through all the other boxes; or made a single large checkmark over the checksheet. Neither action proves anything, but both suggest that the driver did not check each individual item on the list.

It might take the task group member only 5 minutes to conduct this audit, but now she has completed her assignment until the next month and has identified issues that need to be resolved.

The form presented in Figure A-3 can be used by task group leaders to provide a record of which task group members have been given what assignments. This form is followed by a sample completed form in Figure A-4.

GOALS AND OBJECTIVES

Each task group should establish goals and objectives for the next 12-month period. As groups successfully reach each goal, the goals can be replaced with new goals. Task groups should report on their goals and objectives at each FSHC meeting.

It may be helpful to establish goals based on management's priorities. This will give task groups a clearer understanding of what it is that management wants.

MEMBER ROTATION

It is suggested that members (other than chairpersons) of each task group serve for one year and then rotate off to allow other employees the opportunity to serve and get involved in the safety–health system. But rather than rotate all members at once, a staggered system of rotation should be established so that the task groups will not be starting over each year with completely inexperienced members. If half the members are rotated, the remaining experienced members will be able to provide leadership, understanding, and cohesion for the group.

THE FACILITY SAFETY AND HEALTH COMMITTEE

The FSHC is made up of the nine task group chairpersons and is chaired by the ranking manager at the facility. The EHS manager is also a member of the FSHC serving in an advisory capacity. And if the facility has a staff physician, nurse, or other health caregiver, this individual also serves as an advisor to the committee.

The purpose of the Facility Safety and Health Committee is to manage the overall safety and health system.

MEETINGS

Like task groups, the FSHC should meet once each month. This meeting, as described, is to allow department managers who chair a task group a time to briefly report on the most recent meeting of their task group. The FSHC meetings usually last approximately an hour.

A typical meeting would begin with a few opening comments from the plant manager chairing the committee. Each task group chairperson would then, in turn, provide a brief report of what was discussed during the most recent task group meeting. The task group reports would be followed by reports from medical and safety, if need be.

All task group recommendations should be worked out among task group chairpersons in advance of the FSHC meeting at which they will be presented. In so doing, all task groups at this point are in agreement on the recommendation, and it becomes a matter of the plant manager approving or disapproving the recommendation. There is no need for lengthy discussion at this point, and the meeting is not prolonged.

The FSHC meeting is also an appropriate time for the plant manager to delegate responsibilities as needed. For example, a new

OSHA standard might be issued requiring training, inspections, and new rules and procedures. The manager then delegates these responsibilities to the appropriate task groups, who will then ensure that whatever is needed gets accomplished. The Rules and Procedures Task Group will develop whatever rules are needed. The Education and Training Task Group will determine who needs to be trained under the new standard and recommend the best way to get the training accomplished, and so on.

ATTENDANCE

The usual attendance of the FSHC meeting will be those mentioned: the plant manager, each task group chairperson, and safety and medical representatives. It would also be appropriate and helpful for task group members to attend on an as-needed basis.

A member of the Safety Activities Task Group may be responsible for some type of employee recognition program that the task group is planning to implement. It would be helpful for that member—the one most knowledgeable on the issue—to attend and present a short report or simply be there to answer questions that might arise.

Similarly, a task group chairperson might be planning to report on an issue that is of a technical nature. Again, it would be helpful to have the member responsible for that assignment to be present either to give the report instead or answer questions following the report. For this reason, it is suggested that all task group members attend at least one meeting of the FSHC to familiarize themselves with how the committee functions and what the meetings are like.

If a task group or FSHC member is consistently absent and is not a productive member of the group, then he or she should be encouraged to become more effective or replaced.

MINUTES

Just as task groups keep minutes, the FSHC also should keep and distribute accurate minutes of each meeting. Each task group should receive a copy of the minutes of the most recent FSHC meeting.

OTHER PARTICIPANTS

In addition to the task group and FSHC members, there will be ample opportunity for other employees to serve in some capacity in the safety–health effort. These will be described in greater detail in the following chapters, which address specific task groups.

RESPONSIBILITIES

Manager
The ranking manager at the facility

- Holds department heads responsible and accountable for their line management safety and health responsibilities
- Ensures that line management incorporates safety and health into their routine activities
- Ensures that line management routinely enforces safety and health rules and procedures
- Chairs the monthly Facility Safety and Health Committee (FSHC)
- Appoints the task group chairpersons
- Holds task group chairpersons accountable for their assignments

Department Head
The facility department head

- Holds assigned supervisors responsible and accountable for their line management safety and health responsibilities
- Ensures that line supervisors incorporate safety into all their routine activities
- Routinely enforces safety and health rules and procedures
- Serves as a member of the FSHC
- Chairs a task group
- Holds task group members responsible for their assignments

Supervisors
Supervisors

- Hold assigned employees responsible and accountable for their assigned safety and health duties

- Ensure that employees incorporate safety and health in all their assigned jobs
- Routinely enforce safety and health rules and procedures
- Conduct monthly safety meetings with all employees
- Inspect assigned work area monthly
- Report safety inspection results and safety meeting minutes to department head
- Involve employees in work area safety inspections
- Serve on task groups
- Coordinate specific assigned task group activities

Employees
Employees

- Follow applicable safety rules and procedures
- Incorporate safety and health into all jobs
- Promptly report safety defects
- Participate in work area safety inspections
- Attend and participate in monthly safety meetings
- Serve on task groups
- Coordinate specific assigned task group activities

Safety and Health Staff
Safety–health staff

- Advise manager, department heads, supervisors, and employees concerning matters of safety and health
- Audit safety performance throughout the operation
- Keep current and advise management concerning regulations, standards, and codes, as well as state-of-the-art safety and health
- Serve as secretary to the FSHC
- Serve as a resource to all FSHC task groups

SUMMARY

The FSHC, under the direction of the ranking manager, is the heart of the safety–health program. Employee safety and health will rise and fall with the quality of management and leadership provided by this group. For maximum success of the system, three points should be remembered:

1. *Commitment*: A lack of commitment will be quickly sensed by employees. This attitude will then spread throughout the organization, leading to failure of the system. Managers, department heads, and supervisors must maintain a strong commitment to preserving the safety and well-being of employees and demonstrating a high level of commitment through setting a good example and expecting employees to follow that example.

2. *Accountability*: No one can be with all the employees every minute of the day to ensure that they are wearing proper personal protective equipment (PPE), following lockout procedures, or doing all the other things required by the job to be safe. And that is not how it should be. As adult workers, employees must assume responsibility for their own actions, including working safely. Unfortunately, it doesn't always work that way. For their own reasons, employees often violate safety and health rules. Managers and supervisors who fail to hold employees accountable for their actions are not doing their jobs right. Accountability should rank high among the manager's priorities.

3. *Assignments*: The importance of assignments as described in this chapter cannot be overstated. This is how employees get involved in the safety–health process, and when assignments are not made, the system will not be nearly as effective.

5

The Safety Activities
Task Group

When you have a good thing, you want people to know about it. You want to advertise and spread the word. This is, in part, the role of the Safety Activities Task Group—to promote and sell safety and health. Their job is to generate and maintain interest in workplace safety and health and take steps to get employees excited and involved.

PURPOSE

The purpose of the Safety Activities Task Group is to oversee the entire EHS system to ensure that it is functioning properly and that injuries and illnesses are being reduced. They will also coordinate the overall activities involved in promoting safety, health, and environmental issues.

This is the creative group, and there is no limit to the things this group can do in order to achieve their goals. They can organize

---

poster contests, organize employee recognition programs, start or improve a company newsletter, and much more.

Additionally, the Safety Activities Task Group is responsible for overseeing the entire safety–health effort. They must determine whether the program is pertinent to the needs of the organization. It would do little good to focus injury prevention efforts on back injury prevention if the company were not having back injuries. Instead, the Safety Activities Task Group would examine injury and illness data in order to focus efforts in the right direction. If records indicate a high number of cuts, for example, the task group would work to develop special emphasis programs geared toward prevention of such injuries.

The task group should review the entire program annually to ensure that other task groups are also focusing efforts in the appropriate direction. It is usually best to do this in the fall of the year to help develop the safety and health focus for the coming year.

PROGRAM IDEAS

Some ideas that may be used by the task group to ensure program effectiveness include the following:

- Review the entire EHS system annually (in October, November, or December).
- Provide and maintain safety-related signs at facility entrances and exits.
- Select and distribute appropriate safety posters.
- Coordinate all safety bulletin boards.
- Recommend an annual safety–health emphasis program.
- Coordinate all contests and awards programs.
- Provide individual awards for employees who have worked for many years without incurring any serious injuries (medical cases).
- Suggest special-emphasis programs to correct specific safety and health problems.
- Hold annual facility safety and health conferences for selected supervisors and employees.
- Distribute special safety and health bulletins to employees on specific items such as holiday safety, vacation safety, driving safety, and home safety.

- Coordinate departmental safety display contests.
- Recommend specific individuals to attend special conferences.
- Arrange for individuals to visit other facilities and review their safety programs.

SUGGESTED ACTIVITIES

The overall activities of the Safety Activities Task Group involve identifying actual or potential safety and health problems and recommending programs, plans, and activities to improve safety, increase supervisory and employee safety and health awareness, and solicit ideas and suggestions from supervisors and employees. Some of these activities are:

- Routinely review all safety and health statistics to determine the types of injuries and illnesses that are occurring and recommend programs for improvement.
- Solicit suggestions from other task groups concerning particular problems that are occurring that need special emphasis by the Safety Activities Task Group.
- Coordinate all safety contests and awards programs. Consider programs such as
   Individual safety awards for on- and off-the-job injury prevention;
   Section and area safety awards for good safety performance;
   Division safety awards for working at least a million hours without injury resulting in time lost from work.
- Develop an annual safety emphasis program aimed at solving particular safety and health problems.
- The Safety Activities Task Group is responsible for coordinating all safety posters, signs, and general publicity concerning safety activities. One consideration would be to develop a safety display board, which would be placed at the main employee entrances, so that employees can see the daily progress being made toward injury and illness prevention.
- This task group would coordinate all contests pertaining to on/off-the-job safety hazards such as driving safety, vacation safety, holiday safety, and hazards associated with certain

seasons of the year such as winter driving conditions, boating accidents, and other home and recreational safety needs.

MEETINGS

When the Safety Activities Task Group meets each month, topics of conversation should center on how the overall EHS system can be improved. Member assignments will be related to those programs within the system that are intended to recognize efforts, broadcast safety and health, and generally ensure that employees are constantly reminded to work safely. Figure 5-1 is a suggested list of monthly topics that can be used as is or altered to fit the needs of the organization. Each topic listed also can be used as member assignments.

MEMBERSHIP

The Safety Activities Task Group should have at least one representative from each major section. Each member would represent his/her particular section regarding activities and programs. It is, therefore, necessary for each member to keep in touch with all section groups to ensure that proper consideration is being given to each review item.

PROGRAM REVIEW

The following should be reviewed by members of the Safety Activities Task Group in an effort to evaluate the effectiveness of the facility safety–health program:

- How are safety and health items communicated to employees?
- Are employees well informed concerning the organization's and work area safety program?
- Do employees feel that they are a part of the safety–health program?
- Are supervisory safety and health efforts aimed toward solving work area safety and health problems?

Task Group Monthly Review Items				Page 1 of 1 Date:	

This sheet is for use by Facility Safety and Health Committee (FSHC) Task Groups in establishing an annual review schedule to ensure that those items needing review are properly reviewed and reported to the FSHC.

Task Group: Safety Activities		Chairperson:		Year:	
Month	**Review Item**	**Reference**	**Assigned To**	**Comments/ Suggestions**	**Date Completed**
January	Annual safety and health program presentations				
February	Safety meetings				
March	Annual program update				
April	Incentive programs				
May	Safety bulletins				
June	Annual program update				
July	Safety signs				
August	Safety contests				
September	Awards				
October	Annual program review				
November	Program review findings				
December	Proposed annual program				
Other Special Assignments:					
Other Comments:					

Figure 5-1. Task group monthly review items.

- How are employees involved in work area safety and health efforts?
- Are individual employees recognized for their individual safety and health efforts?

- Are entire work groups recognized for their safety and health efforts?
- Does the work area have a specific safety and health bulletin board?
- Do employees attend safety meetings?
- Do employees submit safety and health suggestions?
- What do employees believe should be done to improve work area safety and health?

COMMUNICATION

The Safety Activities Task Group should draw upon the many resources available to them, including soliciting opinions from other employees on what improvements or changes could be made in the safety–health program and information that would be of interest to fellow employees.

NEWSLETTERS

These same resources can provide fodder for a company newsletter. Some employees have a knack and desire to write and will be more than willing to serve as scribes for this effort. Other employees, while they may not have an interest in providing copy for the newsletter, will be glad to contribute information. Getting employees to contribute to a newsletter is a great way to get employees involved. In addition to providing information for safety-related articles, employees are usually anxious to share in the events of the family. We take pride in our children's graduations, the birth of a child or grandchild, employee birthdays, and other milestones. Such information makes good newsletter fodder.

COMMUNITY BROADCAST

There is no reason to keep successes hidden from the public. When good things are happening in the workplace, it pays to get it out to the public. Every employer should want to be a good corporate neighbor. We want to be a thriving and accepted part of the community, and numerous things can be done to both promote employee achievement and enhance the company image.

Local radio and television stations are always seeking public service announcements. Also, newspapers are looking for copy. This is free advertisement for the company's good will! A picture of the task groups along with an article describing who they are and what they do toward protecting workers is one of many ways to get word out to the community. After all, the employees they are protecting through their efforts are not only uncles, aunts, cousins, and friends; they are also the same people who frequent the local malls and movie theaters. Local citizens can certainly appreciate the injury prevention efforts of task group members.

PROFESSIONAL JOURNALS AND ORGANIZATIONS

Some successes warrant even greater exposure. Many of the professional safety and health publications print stories of company successes in safety and health. These should be given consideration when promoting the safety and health efforts of the company. The staff safety and health manager should be able to provide information to the Safety Activities Task Group on various magazines and journals.

SIGNS

Another of the many tools that this task group would use to promote safety and health would be signs. Signs do not make employees safe, nor do they cause them to work safely. But as the safety-conscious culture continues to grow throughout the organization, the signs can serve as a reminder to workers of the importance of safety at the facility. They can serve as a daily reminder that management is concerned with employee safety.

This task group can determine what signs may be appropriate to post at entrances and exits as well as throughout the facility.

POSTERS

Another good source of communication is posters. Posters, like signs, will not make employees safe. But the creation of posters with the hope of winning a poster contest and having the poster published and posted throughout the facility will go far in increasing the employee's awareness level.

Better still, a poster contest for the children or grandchildren of employees is a great way to generate interest and at the same time get some good inexpensive advertisement for safety and health.

BULLETIN BOARDS

Dedicate a bulletin board or bulletin boards to publicizing safety. Have a safety bulletin board in all areas where employees may congregate such as by main exits, lunch rooms, time clocks, and water coolers. Use them to display matters of safety-related interest such as contest rules, progress, and winners; pictures of task group members; and hazards identified in the workplace. A member of the Safety Activities Task Group would maintain all safety bulletin boards.

While Safety Activities Task Group members would have communication as their assignment, they would have other volunteer employees to help put it all together. These same volunteer employees would likely serve as members of the Safety Activities Task Group at some point in the future.

INCENTIVE PROGRAMS

Another tool used by many employers to increase employee involvement is the incentive program. But a word of warning would be in order. Although incentive programs can and often are very beneficial, they can also be detrimental to the safety–health program if not managed properly.

It is important not to let the incentive program get out of hand. If the value of incentive awards is significant, then the risk of not reporting injuries and illnesses increases. Peer pressure can become so strong that employees, out of fear of reprisal from fellow workers, either will not report injuries at all or may lie about how the injury occurred.

It is advisable to keep rewards small. It is possible to get as much mileage out of company or group pride as it is a $500.00 savings bond, or a trip to Hawaii. The choice parking spot in the employee or the executive parking lot can be awarded to the most safety-conscious employee of the month. Rewards do not need to be expensive; nor should they create a frenzy to win them.

It should not be the intent of incentives to "buy" employee's safe behavior. That is a requirement that goes with the job. Incentive programs, if used at all, should be used simply as a tool to help get employees involved and improve awareness levels.

The opportunities and ideas for increasing and sustaining employee involvement are endless, yet everything that the Safety Activities Task Group does, even the little things through small incentives, can pay off in big dividends and contribute in a great way toward building the safety culture.

ANNUAL EMPHASIS PROGRAMS

An annual emphasis program allows employers to focus injury prevention efforts on those things that need attention rather than waste efforts on issues that are of little consequence at the time. Through evaluation of injury and illness records, along with accident reports and investigations, the Safety Activities Task Group can identify the types of incidents that are occurring and the resulting injuries. With this knowledge they are now capable of developing programs that target prevention of targeted incidents. If, for example, the evaluation reveals a high incidence of strains, the emphasis program for next year can be on strain prevention. Now contests, posters, incentive rewards, and other strategies can address prevention of various types of strains. Or, specific types of strains can be focused if need be.

In similar fashion, employers can develop annual emphasis programs to target anticipated issues. Maybe there is growing concern over industrial espionage. The Safety Activities Task Group, working along with the Security Task Group, can develop an annual emphasis program to target security and protecting trade secrets.

A member of the Safety Activities Task Group would be assigned annual emphasis programs and would coordinate the program. To help with this effort, they could solicit information from the Incident Investigations and Inspections and Audits task groups. This same information can be shared with the rest of the Safety Activities Task Group to use when conducting the annual system review and plan development.

AWARDS AND RECOGNITION

Everyone should be recognized for a job well done. And while we are usually quick to recognize poor performance, we aren't always as quick to recognize and reward good safety performance. Personal recognition is an excellent way to get employees to buy into safety and health.

Not only should we recognize good personal performance; we should also reward the good performance of departments, sections, or work groups. This helps build camaraderie and teamwork.

As with incentives, rewards don't have to be great. Simple recognition in the newsletter, posting an employee's or group's picture in a prominent location, or a free lunch in the company lunch room can do wonders for employee morale and help the safety–health culture grow.

A member of the Safety Activities Task Group would be assigned the responsibility of awards and recognition and would coordinate with the help of volunteer employees to create recognition programs. This group could also work with the task group member responsible for publicity to assure the recognition is properly broadcast to others.

CONTESTS

Contests can be an effective and fun way to bring employees into the safety and health system. Like all other activities, there is no limit to what and how contests can be organized.

Poster contests, as mentioned earlier, are favorites, especially for children and grandchildren of employees. Slogan contests, safety bingo, and safety Jeopardy—the list goes on as to what is available or what can be created to suit the employee population.

It is important to provide all employees a chance to win. Contests based on skill and literacy can eliminate some employees who may feel that they have no artistic talent. For this reason it is advisable to sponsor contests of different types. In addition to poster or art-related contests, there could also be contests based on trivia or safety knowledge.

The effectiveness of drawings is usually limited. Drawings limit the number of winners and are not based on any effort of the participants. Further, when an employee fails to qualify for a drawing because of an accident or injury, that employee no longer has an incentive to continue to work safely.

Contests based on team participation are often more effective. The following should be considered when developing contests:

- Offer an opportunity for all employees to win.
- Offer meaningful challenges in accident prevention.
- Use balanced groups with no built in advantage over other groups.
- Provide annual awards or million-work-hour awards for each member of the group.
- Keep awards simple and inexpensive rather than an expensive award for only a few lucky winners.
- Provide tangible awards that employees can keep rather than money, which will soon be gone, along with the reason why the employee earned it in the first place.
- Publicize the contests and awards to all employees.
- Keep participants informed of the status during the contest period.
- Award prizes in a timely manner.

MOTIVATION

Motivating employees is certainly a large part of the responsibilities of the Safety Activities Task Group. There are different ways to motivate people. One can be motivated, for example, by fear. For instance, when the supervisor tells the employee that he will be fired if he is late for work again, the employee will make a special effort to get to work on time, unless the job doesn't mean anything to him. Likewise, if employees watch a video that graphically depicts eye injuries, the employees will be diligent in wearing eye and face protection—for a while. But herein lies the drawback of using fear as a motivator. Sooner or later the images seen in the video begin to fade and employees resume their poor habits of not wearing proper eye protection. Fear is not a longlasting motivator.

People are often motivated by desire. When we offer employees something in return for following safety rules and procedures, the level of response will be proportionate to the level of desire for the reward. If employees are told that they will be given jackets, concert tickets, money, or other perks for not having a recordable injury, eventually something will come along that holds greater appeal than their willingness to follow all the rules. Again, employees will revert to not following rules. They will begin to take shortcuts and make a conscious decision not to lock out equipment so that they can finish the job earlier, or choose to violate other rules that inconvenience them.

So, while fear and desire can be strong motivators, they are not long-term motivators. If we want people to become motivated and stay motivated, we must convince them that following safety and health rules is really in their best interest and the benefits of compliance hold a greater reward than does nocompliance.

The way to achieve this long-term motivation is to make employees part of the safety–health process, to get them involved in some way. The Safety Activities Task Group can do a lot to get employees involved, well beyond the small number of employees who are serving as task group members.

When employees begin to realize that it is their safety–health system and not that of the EHS manager, then compliance becomes more of a personal matter. They no longer look at the EHS manager as the safety cop, but rather as a fellow employee with the job of coordinating the safety–health effort and serving as a resource. They no longer see the safety and health rules as something shoved down their throats but as a part of doing the job with inherent value.

SYSTEM REVIEW AND PLANNING

As stated earlier, one of the primary functions of the Safety Activities Task Group is to oversee the entire EHS system. The task group is responsible for coordinating safety and health to ensure that the system is meeting the needs of the organization. When injuries, illnesses, and accidents are increasing, this indicates that the overall safety awareness of the organization is decreasing.

The Safety Activities Task Group must stay apprised of incidents that are occurring in order to keep the system moving in the right direction and address the safety and health needs of the organization. To do this, the task group should closely follow incidents by getting reports from the Incident Investigation Task Group. This will let them know what incidents are occurring, and whether they are serious or minor in nature.

Similarly, a monthly report from the Inspections and Audits Task Group will let them know when problems keep recurring or are not being corrected. This will allow the task group to focus attention on those problems and their causes in order to create prevention programs.

The only way for the EHS system to be effective is to constantly improve safety awareness of the line organization. Employees get their safety awareness from supervisors' safety awareness. Supervisors' safety awareness is a result of department head safety awareness. And the department heads get their safety awareness from that of the facility manager. Thus, employee safety awareness is a result of the ranking manager's awareness level and will reflect his/her safety commitment and leadership.

The Safety Activities Task Group should review the facility EHS system at the end of the year. Following this review they should be able to identify direction for the system in the coming year and plan accordingly. A thorough review will reveal weaknesses within the system that can then be addressed by planning activities that will focus on these weaknesses.

Each task group should provide the Safety Activities Task Group with a list of three or four goals and objectives that will then be worked into the annual plan. The plan is implemented on the first day of the year and reviewed periodically and maintained by the Safety Activities Task Group.

Once each year, the facility manager (FSHC chairperson) should complete an evaluation of the EHS system and make recommendations to the entire FSHC. A Task Group Evaluation Form is available in Figure A-5 for this purpose.

ANNUAL SAFETY–HEALTH CONFERENCE

Many employers sponsor annual safety–health conferences or will send selected employees to safety conferences as a reward for their participation in the EHS system. An annual safety–health conference can be a good way to cap off or reward a successful year. A conference can be a small affair intended only to give members of the FSHC organization an opportunity to review the current safety–health system and make suggestions for improvement. Or, it can be a larger affair, inviting guest speakers from among consultants, OSHA, the company's insurance carrier or local professionals from fire, police, and the medical community. The latter can also serve to meet the required safety and health training needs of workers.

A member of the Safety Activities Task Group would be assigned conferences and would coordinate the conference. Other employees could be assigned to assist with conference planning.

If planning a conference, the following schedule is suggested:

Month	Activity
February	Appoint conference coordinator
March	Select conference date
April	Confirm meeting room arrangements
May	Approve agenda
June	Notify speakers
July	Confirm speaker preparation
July	Arrange conference details
September	Hold conference
October	Prepare conference report
October	Distribute conference report

A program review conference would consist of each task group chairperson giving a brief report on the activities of his/her task group followed by a discussion. Someone should be assigned to take notes and prepare a report afterward for distribution to all supervisors and attendees.

In addition, a company safety and health picnic could be held for employees and their families. Following are some suggestions for sponsoring safety and heath picnics:

- If outdoors, schedule the picnic when the weather will be favorable.
- If possible, schedule gatherings at a time when the largest number of employees can participate. If possible, shut down the operations to enable all employees to participate, or schedule on a planned downtime day.
- Provide activities for all ages.
- Remember it is a safety and health gathering. Include activities that are safety- and health-related.
- Invite guest speakers; local and corporate dignitaries; entertainment, safety, and health equipment distributors; and so on.
- A safety–health picnic is also a good time to present awards and recognize safety efforts of groups and individuals.

6

The Rules and Procedures Task Group

Rules and procedures are an essential part of any organization. Consider professional football, for example. The rules of the game have evolved over the years to make the game better, and also to make it safer. Injuries in the game today are fewer and less severe because of improvements in equipment, but to no less degree because of changes in the rules. The same can be said for the workplace. Rules have been established to make work safer for employees. While some employers may consider OSHA law to be a costly intrusion and a deterrent to free enterprise, the amount of money saved by employers as a result of safety enforcement is immeasurable. There is no question that the laws have had a positive impact in reducing deaths and injuries in the workplace.

The OSHAct was enacted by Congress out of recognition of a high rate of injuries and fatalities and the need to do something about it. The rules that followed came initially from numerous other sources that were incorporated by reference. These rules were initially created as consensus standards and make up the barest minima for protecting employees. Many of the standards in existence require employers to have written programs with additional written procedures. The safety and health manual has continued to increase in size since 1970 and the Rules and Procedures Task Group plays a major role in ensuring that the needed rules at the facility are developed and maintained.

In addition to the rules and procedures required by various OSHA standards, each employer should have a set of general rules and procedures that apply plantwide. The role of the Rules and Procedures Task Group is twofold: to coordinate the writing of rules that need to be written, either general safety and health rules that apply plantwide or rules and procedures specific to certain areas or tasks. Usually this is less of an undertaking than one might think since most employers already have rules established.

To write some rules requires not only specialized understanding of what is required by law but also perhaps an understanding of equipment and processes. For this reason it is beneficial to have an ad hoc committee made up of employees and supervisors who write procedures such as lockout procedures. This not only increases the size of the pool of experts; it also gets more people involved in safety and health and allows the employer to direct talent to where it is needed and best used. In some circumstances it may be necessary to have the expertise of a safety professional or industrial hygienist to assist with the development of programs and policies.

In addition to coordinating the writing of rules, the Rules and Procedures Task Group is to review existing rules annually to ensure that the rules have not become obsolete or present problems for employees or the employer. Many rules have become obsolete over the years simply because equipment has also become obsolete or technology has changed, rendering many rules useless.

A good example is the many municipalities around the nation that still have local ordinances banning drinking water from the horse trough. The City of Atlanta still has a law that prohibits anyone from tying a giraffe to a lamp post. For whatever reason, rules become obsolete or in some cases they were poorly devised and do not permit employees to do their jobs and obey the rules at the same time.

A fiberglass insulation manufacturer had a 24-in. blade that would move out across a conveyor belt to cut insulation on the belt.

To protect workers in the area, a guard was constructed around the blade. The guard was made of expanded metal and was 7 ft high with a built-in gate for employee access, mostly for maintenance personnel.

An employee was observed standing inside the guard with the gate open as she turned an adjustment screw with a screwdriver. When the safety and health manager saw her he ordered her out, asking why she was inside the guard. "My job requires me to adjust this screw several times a day, and you can do it only when the machine is running. But no one ever asked me about my job," she went on, "and I don't have a screwdriver 6 ft long that allows me to stand outside the guard and do my job."

Clearly, the employer had created a situation in which the employee was unable to do her job and comply with the rules at the same time. It now was necessary to go back and reexamine the situation and redesign the guard.

The Rules and Procedures Task Group will review rules annually to eliminate any that have become obsolete and to make changes in rules where necessary.

PURPOSE

The purpose of the Rules and Procedures Task Group is to ensure that all safety and health rules and procedures are established, communicated, and effectively maintained, and to make sure that general safety and health rules and procedures are compatible with section safety rules and procedures.

MEETINGS

What assurance does the employer have that all required safety and health rules are in place and updated as needed? Are existing rules and procedures reviewed annually to ensure that they are still appropriate or needed changes are identified? At monthly meetings of the Rules and Procedures Task Group, group members review rules and procedures related to specific subjects to ensure continued compliance. Figure 6-1 provides a suggested list of topics that can be reviewed each month by the group. Each monthly topic also may serve as a member assignment.

	Task Group Monthly Review Items			Page 1 of 1 Date:

This sheet is for use by Facility Safety and Health Committee (FSHC) Task Groups in establishing an annual review schedule to ensure that those items needing review are properly reviewed and reported to the FSHC.

Task Group: Rules and Procedures   **Chairperson:**   **Year:**

Month	Review Item	Reference	Assigned To	Comments/ Suggestions	Date Completed
January	Injury /illness recordkeeping				
February	General rules				
March	Lockout/tagout				
April	Confined space entry				
May	Personal protective equipment				
June	Fall protection				
July	Barricades				
August	Cutting devices				
September	Materials handling				
October	Recordkeeping (other than injury/illness)				
November	Material storage				
December	Visitors and contractors				

Other Special Assignments:

Other Comments:

Figure 6-1. Task group monthly review items.

RULES REVIEW

Rules and Procedures Task Group members should review the general safety and health rules each year. Each member should be assigned specific rules for individual review. The following procedure is suggested for rules review by task group members:

- Review the rules.
- Discuss rules with supervisors and employees.
- Audit the workplace to determine whether rules are being followed.
- Recommend to the Rules and Procedures Task Group any needed rule revisions, training, or enforcement.

Discussion of rules takes place during task group meetings, and any changes should be recorded in meeting minutes.

Any suggested changes should be submitted to all other task group chairpersons for their input and discussion before they are presented to the FSHC. Once the rules and suggested changes have been tweaked and agreed on by other task group chairpersons, they should be submitted at the FSHC meeting. Any major changes should be accompanied by an explanation as to why such change is being recommended.

On approval of rules or rules changes, it now becomes necessary to communicate the information to employees and supervisors. At this point the Education and Training Task Group would coordinate training programs to ensure that all affected employees and supervisors are made aware of any changes.

The following list should serve as a guide in examining existing rules or creating new ones:

- Are rules stated in writing?
- Are rules and procedures current?
- Are rules and procedures available?
- Do supervisors know and understand the rules?
- Do employees know and understand the rules, and are they able to follow them?
- Do employees follow the rules? If not, why not?
- Do supervisors enforce the rules? If not, why not?
- Are general rules and procedures contrary to other rules?

GENERAL SAFETY AND HEALTH RULES

Most employers have a list of general safety and health rules that apply plantwide.

Following is a typical list of general safety and health rules:

1. Report all injuries immediately to supervision or the medical staff.

2. Horseplay, scuffling, and fighting are prohibited.

3. Unauthorized operation or maintenance of equipment is prohibited.

4. Rings, bracelets, wristwatches, loose adornments or garments, or long hair (below the shoulders) must not be worn within 3 ft of operating machinery or moving product.

5. All guards and safety devices must be in place before equipment is operated, except as provided in written approved procedures.

6. Repairing, adjusting, servicing, or cleaning machines and equipment that will expose employees to hazardous energy sources can be conducted only by trained and authorized employees using "lock, tag, and try" as required in specific procedures.

7. Roped-off and barricaded areas can be entered only with permission of personnel working in the enclosed area or of the supervisor responsible for the work.

8. Approved safety glasses must be worn as minimum eye protection for protection against flying objects, glare, and radiation as specified by job procedures. Goggles are required for employees using air hoses to blow off equipment and for liquid splash protection.

9. Strike-anywhere matches, firearms, explosives, and ammunition are prohibited unless authorized.

10. Smoking is permitted only in designated areas. Cigarettes, cigars, and matches must be discarded only in ashtrays and specifically identified containers.

11. Posted speed limits must be observed by drivers of both on-site and outside vehicles.

12. Intoxicants, narcotics, and persons under the influence of these substances are prohibited in the plant, except as authorized by medical staff.

13. Access to emergency exits, evacuation routes, and emergency equipment must not be obstructed.

14. Running on site is prohibited, except to prevent loss of life or serious injury.

15. All chemicals used, purchased, or accepted must be approved through the Facility Safety and Health Committee, and chemical containers must be properly identified in accordance with company policy regarding container labeling.

16. All scissors, knives, razor blades, and sharp-pointed tools used by employees must be specifically approved by supervision for their intended use.

SAFETY AND HEALTH MANUAL

The safety and health manual catalogs the organization's rules and procedures and serves as the graphic representation of the organization's safety and health system. It is the bible on matters of safety and health and should be available to all employees. For this reason there should be a copy of the safety and health manual in all departments and operating areas, with easy access by employees at all hours.

The Rules and Procedures Task Group is responsible for maintaining the safety and health manual. A member of the task group should serve as custodian and be assigned to coordinate upkeep of the manual.

Each manual should be identified by title, number, or in some way that distinguishes it from other manuals. An audit of each manual is conducted at least annually by the assigned task group member to ensure that each manual is being properly maintained. A manual checklist and a safety–health manual audit sheet are provided in Figures A-6 and A-7, respectively.

Any changes in the rules should be sent to area manual custodians for inclusion in their manuals, replacing obsolete pages with the new pages containing any changes. A system of checks and balances should be established to ensure that changes were implemented and all manuals are current following distribution of new information. It is recommended that each custodian be required to return any obsolete pages to the safety and health manual coordinator as notification the changes have been made.

SAFETY AND HEALTH RECORDKEEPING

Numerous records are required by various entities, including OSHA, workers' compensation, and the Environmental Protection Agency (EPA). The Rules and Procedures Task Group should be responsible for reviewing, evaluating, and recommending improvements for safety and health records. At least two members of the

Rules and Procedures Task Group should conduct this review at the beginning of each year.

TASK GROUP COMMUNICATION

When incidents occur, regardless of injury, they often result from a violation of rules or procedures. Violations may occur because a rule is hard to follow, may not be known, or is not enforced, or there may be simply a conscious decision on the part of the offending employee to violate it. Regardless, an effort should be made to determine the reason for the rules infraction. The Incident Investigations Task Group should forward copies of all investigation reports to the Rules and Procedures Task Group for review to determine whether changes need to be made in the rules.

Similarly, inspections and audits may repeatedly reveal a rules violation of some type. The Inspections and Audits Task Group should send inspection findings to the Rules and Procedures Task Group for review to determine whether the rules, as they presently exist, make compliance difficult.

An employee may be observed not wearing his safety glasses. The rules are reviewed, and an interview with the employee reveals that he is unable to see to work safely because his safety glasses fog up on hot days. Now, rather than change the rule, the Rules and Procedures Task Group would inform the Health and Environment Task Group, since this group is responsible for personal protective equipment. The Health and Environment Task Group can then get antifog safety glasses, not only for this but for all employees. This will now allow them to perform their jobs safely while complying with the rules on eye protection.

7

The Education and Training Task Group

Every employee within the organization requires some amount of annual training in safety, health, security, and/or environmental protection. The type and amount of training required depends entirely on their jobs. Ensuring that all employees receive the required training is a huge undertaking. This is the role of the Education and Training Task Group.

Many larger companies have training managers with a staff to assist in scheduling and conducting training. Smaller facilities, however, rarely have this luxury. Instead, they rely on the staff safety–health professional, who may have other collateral duties, to ensure that all employees receive the required training. Figure 7-1 is intended to help the Education and Training Task Group manage the required training. Suggested topics that should be covered during monthly meetings are listed. Each month a single topic from the list is the major topic for that month. The topics listed can also serve as member assignments.

Ensuring that all employees receive the training required by law can be a daunting task. This is where the Education and Training

Task Group Monthly Review Items				Page 1 of 1 Date:	

This sheet is for use by Facility Safety and Health Committee (FSHC) Task Groups in establishing an annual review schedule to ensure that those items needing review are properly reviewed and reported to the FSHC.

Task Group: Education and Training

Chairperson:

Year:

Month	Review Item	Reference	Assigned To	Comments/ Suggestions	Date Completed
January	Annual training schedule review				
February	New employees				
March	Transferred employees				
April	New equipment/machines				
May	New procedures				
June	Access to employee exposure and medical records				
July	Safety meetings				
August	Manager/supervisory training				
September	Visitor/contractor training				
October	Job safety analysis				
November	Training needs revealed by accident investigation				
December	Employee feedback				

Other Special Assignments:

Other Comments:

Figure 7-1. Task group monthly review items.

Task Group comes in. This group determines who needs what training and then makes all necessary arrangements to get it done. In other words, they manage training. Members of the Education and Training Task Group are assigned specific issues for which they will determine training requirements and to whom those requirements apply.

REQUIRED TRAINING

Many OSHA and consensus standards clearly indicate training requirements for those employees who will be using equipment or performing a task. Some regulations merely imply training requirements, making little or no specific reference to the subject of training. For example, the OSHA emergency eyewash and shower standard found at 1910.151(c) states that eyewashes and showers "shall be provided within the work area for immediate emergency use." Nowhere does the standard make mention of training requirements. The word "training," in fact, will not be found in the eyewash standard. Yet, it does little good to provide the needed eyewash and shower if the affected employee does not know how to use it; therefore, there is an implied requirement to train those who may work with injurious materials and may one day have need of an eyewash.

The Education and Training Task Group, along with the help of the safety–health professional, must ferret out such training requirements to ensure that all employees who need the training get it. A thorough list of OSHA required training can be found in Figure A-8.

NEW AND TRANSFERRED EMPLOYEES

Statistics show that new employees and employees who are newly transferred are likely to suffer more incidents than are the veterans who have been around for a while. Such statistics make perfect sense. The new employee as well as the transferred employee is working in a different environment, surrounded by different equipment or people, and learning and practicing new procedures. Whenever things are different for employees, the likelihood of an incident increases dramatically.

For this reason, special attention should be given to identifying and training employees who are new or working in a different environment.

TRAINING QUALITY

A poorly trained employee will not have the necessary knowledge and skills to perform tasks safely. It is this lack of knowledge and skill that can lead to injury or death. For this reason the employer wants to make every effort to ensure that employees receive quality training, as training effectiveness is a direct result of the quality of the training. Hence, the Education and Training Task Group is responsible for training quality. Task group members should be assigned to monitor training and collect information in order to enable them to evaluate training effectiveness.

A task group member should also determine whether training methods are adequate. Long gone are the days of chalkboards and overhead projectors as our primary training media, with the slide projector not far behind, in the race for obsolescence. Digital photography and computers have given us much more versatility along with improved quality. As we continue to see advances in technology, our training tools continue to improve, offering even greater versatility and effectiveness, resulting in more interesting training. Employers should strive to remain on the cutting edge of training technology.

Instructors should make training as interactive as the subject will allow. Statistics show that the greater the participation of trainees during the training process, the more effective the learning experience. Retention rates increase significantly as trainees get more actively involved in the training through reading, hands on, simulation, discussion, and other procedures. Training methods affect training quality and should be given consideration.

INSTRUCTOR SELECTION

A key issue in training quality is instructor selection. A weak instructor will not deliver effective training. Likewise, an instructor who does not possess the requisite knowledge will not deliver satisfactory results.

In selecting instructors, there are several things to consider:

- Does the prospective instructor have good experience?
- Does the instructor possess the knowledge needed?
- Is the instructor a good presenter?

It isn't enough simply to have experience and knowledge; an instructor must also be capable of delivering information in a manner that can be understood by employees and with interesting presentation. If the instructor speaks above the heads of the trainees, then the training will be lost. Employees will quickly become bored and turn the instructor off. Likewise, an instructor should not be condescending, rude, vulgar, or even suggestively off color. Special care must be exercised to select an instructor who is professional and can deliver the desired training.

The Education and Training Task Group should work with the safety–health professional or the training department to select good instructors capable of bringing the training that will meet the needs of the employer.

DOCUMENTATION

While good training can help employees perform their jobs right, if the training has not been documented, then the employer is unable to determine who needs recurrent training, or when recurrent training is due. Additionally, there would be no record to prove to OSHA that employees were trained. There would be no records to substantiate employee training should an incident occur and wind up in litigation. Training records are vital for many reasons, and the employer must ensure that good records of employee training are properly maintained.

In more recent OSHA standards OSHA has been clear regarding elements that should be included in documentation. Training documentation should include the following:

- Employee's name
- Date and time of instruction
- Subject of training
- Length of training
- Contents of training
- Elements of training (hands-on, classroom, etc.)
- Instructor's name and signature
- Method of verification of skills and knowledge
- Results of verification (test results, peer review, evaluation, etc.)

Records should be understandable and easily accessed. Unobtainable records are of little value to anyone.

The Education and Training Task Group should assist the training department or the safety–health manager in this regard, to ensure that good records of employee training are maintained.

MEETINGS

Meetings are a form of training and therefore should function smoothly. A poorly run meeting will yield poor results and do little more than waste time of those in attendance. For meetings to be effective, there should be an agenda to keep the meeting running on schedule and staying on the subject at hand.

A member of the Education and Training Task Group should be assigned the responsibility of monitoring meetings and meeting effectiveness. This task group member would review meeting schedules, minutes, meeting effectiveness, and follow-up. A Meeting Audit Form is provided for this purpose in Figure A-9.

8

The Inspections and Audits Task Group

Throughout the workplace there are hundreds, often thousands, of things that require inspection to ensure employee safety, and to meet OSHA compliance. Machine guards alone present an overwhelming challenge requiring inspection every 60 days [1910.219(p)]. Fire extinguishers, ladders, forklifts, handtools, electrical tools, and equipment are just a sampling of the many things that require inspection on a regular basis. It stands to reason, therefore, that conducting inspections is too big a job to expect the safety and health professionals, the maintenance department, or even an inspections team to accomplish.

Because of the abundance of inspections required, it is necessary to increase the number of inspectors, and what better way to do this than to get employees involved in the inspections effort. There are two distinct advantages to having employees conduct inspections:

1. It is a way to get all the required inspections accomplished. By including employees, an employer can create an army

---

Effective Environmental, Health, and Safety Management Using the Team Approach, by Bill Taylor
Copyright © 2005 John Wiley & Sons, Inc.

of inspectors, thereby having enough inspectors to meet inspection requirements.

2. Having employees conduct inspections and holding them accountable, as a part of doing their jobs, is a way to increase employee involvement and thus increase awareness in safety and health, in general.

The Inspections and Audits Task Group is charged with ensuring that all inspections are conducted so that equipment and facilities will be in safe condition. The Inspections and Audits Task Group does not do inspections. That bears repeating. The Inspections and Audits Task Group does not do inspections. Rather, line employees and supervisors should be assigned the responsibility for conducting inspections and be held accountable.

As was mentioned at the outset of this chapter, too many things require inspection to expect a small team to be able to complete all the inspections. And even if this were possible, the effort would deprive employees of that needed involvement. There is great value in having employees conduct inspections.

PURPOSE

The purpose of the Inspections and Audits Task Group is to ensure that all inspections are accomplished and documented. In order to manage inspections the task group must first determine exactly what requires inspection. To make this job easier, an OSHA required inspections checklist is presented in Figure A-10.

PURPOSE OF INSPECTIONS AND AUDITS

Inspections and audits are necessary for two reasons. First and foremost, we must maintain a safe and healthful workplace. We must examine equipment, tools, work areas, and other facilities to identify unsafe conditions that may present hazards to workers. It is also necessary to check equipment such as fire extinguishers, eyewashes, and emergency stops, to ensure that they will function properly when needed. In addition to identifying unsafe conditions, adequate inspections and audits will also reveal unsafe activities. They will reveal the actions committed by fellow workers that are not safe, enabling corrective action through additional training or disciplinary actions, if needed.

The second reason for conducting inspections is simply to meet OSHA mandates. However, it is often necessary to conduct inspections more frequently than is required by law. For example, ANSI Z358.1, the consensus standard on emergency eyewashes and showers, says to test eyewashes once per week. It is hardly inconceivable that the water supply line to an eyewash could be struck and broken by a forklift. Should this occur on the day following the formal weekly test, it could be another week before it is discovered that the water to the eyewash has been shut off until repairs can be made. Now, if the eyewash is needed during that time period, the employee who may need it will not have a working eyewash. For this reason it is prudent to exceed requirements and have workers who may potentially need the eyewash conduct a quick informal test each day before beginning their shift.

While on the job, employees should be constantly aware of their surroundings. Are machines not guarded? Are there moving vehicles such as forklifts? Are there unidentified or unauthorized persons present? While this seems like a lot to ask of workers, it is easily accomplished, resulting in extraordinary value in terms of increased awareness. Having employees conduct safety, health, environmental, or security inspections will improve the awareness level of employees as a result of increased participation, thus enabling them to easily identify things out of the ordinary without making an effort to do so.

OTHER INSPECTION CONCERNS
HOW FREQUENTLY SHOULD INSPECTIONS BE CONDUCTED?

To some extent, the inspection frequency has already been addressed by OSHA, EPA, and other governing bodies through regulatory actions. For example, OSHA tells us to inspect fire extinguishers monthly. This means that portable extinguishers must be inspected at least once every 30 days, not 31 or 32 days. "Once per month" is not necessarily monthly. Likewise, there are numerous other standards that specifically state inspection frequencies.

Many standards, however, are performance-based, especially when it comes to determining inspection frequency. While OSHA requires employers to inspect ladders, the frequency of inspection is left up to the employer. Terms such as "inspect regularly" or "frequently" enable the employer to determine frequency according to the level of usage. There is no need to inspect a ladder every day if that ladder is used only once each year for special purposes. On the other hand, if a ladder is used daily, then an annual inspection would not be adequate.

Employers must become sufficiently familiar with standards to meet minimum inspection requirements. Yet, it is often necessary to exceed minimum requirements, and OSHA would be the first to tell us this. It is up to employers to develop safety, health, and environmental systems to meet their particular needs. This includes adequate inspections.

WHEN SHOULD INSPECTIONS BE CONDUCTED?

Some things, such as extension cords, drills, forklifts, and cranes, must be inspected by the user before use on each shift. For this reason, all required inspections cannot be conducted during the first shift. It is necessary to conduct inspections around the clock. Other inspections, if conducted only during the first shift, may not reveal existing problems.

A true example is a new employee who was required to work overtime into the second shift in order to complete an extensive equipment repair. The employee approached the second shift supervisor, asking where the electrical disconnect was for a particular piece of equipment that she was about to work on. When asked why she wanted to know, she was informed by the supervisor that it would not be necessary because, "we don't lock out on this shift."

Or, there was the time when the safety manager decided to take a surprise walkthrough at 2:00 am and found a maintenance employee 18 ft above the floor without fall protection, while intoxicated.

If an employer confines inspections to the first shift, or any particular shift, for that matter, there will be issues that will not be identified. Employees take shortcuts, rules and procedures are not enforced by supervisors and managers, and liberties are sometimes taken with the use of drugs and alcohol. Never overlook second and third shifts.

WHAT MUST BE INSPECTED?

Virtually everything in the workplace must be inspected, including handtools, floors, doors, cranes, and personnel lifts. The list of things that require inspection is no shorter than the company's inventory of equipment and facilities. This is further reason to get employees involved in the inspection process.

Throughout OSHA standards, inspections requirements are clearly stated or perhaps simply implied. The law notwithstanding, any piece of equipment can malfunction, can wear out or can present hazards, so every piece of equipment should be inspected.

Employers should inventory everything and establish an inspections program to ensure that nothing is overlooked. Special attention should be placed on the frequency of inspection to ensure that things are inspected as often as need be.

In addition to equipment, it is also useful to inspect outside and around the facility. Parking lots and fences should be inspected. Someone should get up onto the roof occasionally to see what problems might have gone unnoticed. Are there fall hazards that have been overlooked? Is there evidence of unauthorized employees present on the roof?

Ventilation systems, protective equipment, electrical receptacles, and related equipment and traffic flow, both inside and around the facility, should be examined. This becomes especially important as facilities and equipment grow older and as things change.

INSPECTION RECORDS

An important part of the inspection process is keeping accurate records of completed inspections. Documentation is important not only in order to satisfy OSHA that the employer is meeting inspection requirements but also to enable to the employer to effectively audit inspections in-house. Without accurate records, the employer would have no way of knowing what has or has not been inspected. Without accurate records, the employer would not know where to focus hazard abatement efforts and hazardous conditions would continue to exist. Therefore, it is vitally important to maintain accurate and accessible records.

This is something that can be coordinated by the Inspections and Audits Task Group.

MEMBERSHIP

Members for the Inspections and Audits Task Group can come from across the board. For what task group members will be doing, there are no special requirements regarding technological or other expertise.

TYPICAL ASSIGNMENTS

Typical assignments for Inspections and Audits Task Group members can include

- *Departmental inspections:* Some things are unique to departments or areas, making inspections unique. Non–task group members within such departments would do the inspections and then the task group member with this assignment would review the findings.
- *Shift inspections:* There may be tasks performed only during off shifts. These tasks and related equipment also need inspecting and would be conducted similarly to departmental inspections.
- *Ladders:* There should be a formal ladder inspection program whereby all portable, mobile, and fixed ladders are inspected on a regular basis. The task group member who has this assignment would typically review inspection results to determine whether inspections are made as prescribed, damaged ladders have been removed from service, and basically the program is functioning as it should.
- *Forklifts:* The member of the Inspections and Audits Task Group who is assigned forklifts would pull the daily inspection checksheets each month for review. They would check for thoroughness of inspection; was each item on the checklist examined? They would look for evidence that might suggest otherwise, such as a line drawn down through each box rather than individual checks. Are damaged trucks being removed from service and repaired? Findings would be reported to the task group.
- *Electrical:* Opportunities for electrical hazards are plentiful, and the results of an electrical incident are usually serious. Are flexible cords used in lieu of fixed wiring? Do receptacles have adequate tension? Are receptacles wired properly?

Employers should conduct inspections to identify such hazards, and a member of this task group would review the findings on a regular basis.

- *Hand and portable powered tools:* The Electrical Safety Related Work Practices standard requires employees to inspect electric powered tools and extension cords before use on each shift and whenever the equipment is relocated. Additionally, nonpowered handtools such as hammers, chisels, and screwdrivers should be inspected from time to time. Since some tools may be inspected several times each day, each inspection does not have to be documented. Still, it is necessary to ensure that inspections are being conducted. The individual with this assignment would coordinate to ensure that nonpowered tools are randomly inspected and electric powered tools are inspected in accordance with the OSHA standard.

- *Vehicles:* Vehicles receive a state safety inspection annually, but what about the turn signal or headlight that quits working 2 months following the inspection, or the gas leak that occurs? Drivers must be held responsible for the safe condition of the vehicles they operate. This individual coordinates daily vehicle inspections conducted by operators.

- *Cranes and hoists:* A crane is a machine that will move a load vertically and horizontally; therefore, a chain hoist or come-along on a trolley is considered a crane. Employers are required to inspect cranes on initial installation. There are also requirements for frequent inspections (daily to monthly) and periodic inspections (monthly to annually). Inspection records and check sheets should be reviewed by this individual.

- *New equipment/facilities:* Anytime something is new, it represents change. Anytime there is change, there is an increased opportunity for incidents to occur. For this reason, someone should conduct inspections before new equipment and facilities go into operation. The individual assigned this responsibility will coordinate by assigning inspectors to look at whatever might be new.

- *Stairs:* While non–task group members can be assigned the responsibility of inspecting stairs for damaged railing, inoperative lighting, inappropriate storage under stairs, and so on, the task group member with this assignment would assign inspectors and review findings on a periodic basis.

- *Personal protective equipment:* OSHA requires personal protective equipment of various types to be inspected. Someone needs to determine both the type and the frequency of the inspections and who will do these inspections. The task group member reviews the PPE inspection program and findings.
- *Machine guards:* There may well be thousands of machine guards throughout the facility. OSHA requires them to be inspected every 60 days. The task group member assigned to this effort should make sure that all guards are included and should review findings.
- *Inspection program effectiveness:* This effort can be done as a group or can be assigned to a task group member. There should be a review of the overall results of the inspections program to determine whether incidents are occurring because findings were not complete or recommendations were ignored.

Assignments should be commensurate with task group members' interests and experience. The HSE manager serves as the resource for task group members to turn to and gain knowledge about requirements concerning their respective assignments.

SUGGESTED ACTIVITIES

Certain activities with respect to inspections are common throughout the workplace and require coordination. To ensure that the inspection needs of the organization are being met, consideration should be given to the following:

- How effective are inspections? Review injury, illness, near misses, and other records to determine causes of incidents. Are these incidents the result of a failure to inspect, a failure to detect during inspection, or a failure to correct problems that have been identified? A comprehensive causal review can reveal holes in the inspections program.
- Highlight repeat problem items and problem items not properly corrected at the FSHC meetings.
- Arrange for audits of ventilation systems including lab hoods, welding ventilation systems, and battery room ventilation to ensure that the systems operate properly.

- Establish audit procedures for conducting safety and health audits of all new or modified equipment and processes.
- Schedule electrical equipment and system audits to ensure proper grounding, installation, and maintenance. Check to ensure that receptacles are wired properly and older receptacles have adequate tension.
- Perform analyses to determine the cause(s) of repeat items on safety inspections and audits. Suggest ways to eliminate the causes of these items, such as by introducing improved engineering controls, and improved enforcement of rules and procedures.
- Emphasize to operating groups the importance of maintaining equipment properly and of ensuring that safe work practices and procedures are followed. A basic objective of this task group is to promote awareness that equipment users— and not staff or servicing department personnel—are responsible for their equipment.
- If applicable, have one member chair a Process Hazards Analysis Team to conduct process hazards analyses per OSHA standard 29 CFR 1910.119. This team manages the facility's process safety management program.

MONTHLY REVIEWS

At each monthly meeting of the Inspections and Audits Task Group there should be a review of a specific item. Figure 8-1 is intended as a guide for task groups to follow each month. It can be adjusted as needed to accommodate a facility's specific needs. Each topic listed can also serve as task group member assignments.

HAZARD RECOGNITION

Employees often will be completely unaware of hazards within their work areas. There are two primary reasons for this: (1) employees may not always be tuned in and simply do not see what is there, usually because they feel that safety is someone else's responsibility; and (2) employees generally are not aware of what constitutes a hazard.

If a safety culture is to be created, it is most important that employees be made aware that safety is as much their responsi-

Task Group Monthly Review Items					Page 1 of 1 Date:

This sheet is for use by task groups in establishing an annual review schedule to ensure that those items needing review are properly reviewed and reported to the FSHC.

Task Group: Inspections and Audits   **Chairperson:**   **Year:**

Month	Review Item	Reference	Assigned To	Comments/ Suggestions	Date Completed
January	Departmental inspections				
February	Shift inspections				
March	Crane/hoist inspections				
April	Lift trucks				
May	Vehicles				
June	New equipment/facilities				
July	Ladders				
August	Stairs				
September	Personal protective equipment				
October	Machine guards				
November	Hand and portable powered tools				
December	Electrical				

Other Special Assignments:

OTHER COMMENTS:

Figure 8-1. Task group monthly review items.

bility as the safety manager's. Employees must be held accountable for their own actions or inactions, as the case may be. Only when employees see safety as an integral part of their job can the employer go to the next step toward achieving a safety culture.

Once this has been accomplished, each employee should receive hazard awareness training. One of the simplest and most effective ways to do this is to simply walk through the workplace shooting pictures of different hazards and safety/health violations and possible security weaknesses. Even in the safest plants it's not hard to find hazards. When these pictures are shared with employees along with discussion as to why they constitute hazards, awareness levels begin to go up. These are the building blocks of the safety culture.

This increased understanding of hazards along with the increased awareness will better equip employees to conduct inspections. They will now be able to recognize things that they once ignored.

9

The Health and Environment Task Group

Falls and unguarded machinery are not the only things that will hurt employees. Each year employees are overcome by toxic substances released when welding, handling chemicals, or operating equipment with internal-combustion engines. Employees develop conditions such as tendonitis, carpal tunnel syndrome, or bursitis from manual materials handling or ill-conceived workstations. There are numerous things in the workplace that will have a negative affect on the health of employees.

Likewise, there are hazards in our environment that must be controlled. Accidental contamination of elements of our surroundings can have devastating consequences on the environment, employee health, and the surrounding community, not to mention community relations and a positive company image.

Employers must recognize and comply with the various laws regulating spills, releases, and the proper disposition of chemicals, pollutants, runoff, and hazardous waste.

Effective Environmental, Health, and Safety Management Using the Team Approach, by Bill Taylor
Copyright © 2005 John Wiley & Sons, Inc.

These health and environmental issues present challenges for management. The Health and Environment Task Group helps to ensure that these matters are properly managed.

PURPOSE

The purpose of the Health and Environment Task Group is to ensure that matters involving employee health and the health of the environment are properly managed to protect workers from hazards associated with exposure to hazardous substances, noise, and ergonomic stressors.

PROGRAM IDEAS AND SUGGESTED ACTIVITIES

Some of the recommended activities are

- Conducting an occupational health analysis of all operations listing the potential occupational health hazards. This would include such items as dusts, welding fumes, noise for those personnel working in areas where the noise level is above 85 dBA, solvent vapors associated with metal cleaning, plating and painting operations, fumes and vapors associated with equipment cleaning, jobs involving potential oxygen-deficient atmospheres such as entry into tanks and pits, potential skin and eye irritation from exposure to corrosive chemicals, and manual materials handling problems (ergonomics).
- Establishing and maintaining an inventory of all chemicals.
- Establishing procedures for controlling the purchase of chemicals.
- Maintenance of Material Safety Data Sheets (MSDSs) for all chemicals and hazardous substances used throughout the facility.
- Establishing routine monitoring procedures to survey the potential health hazards throughout the facility.
- Routinely auditing to ensure that adequate employee protection is provided concerning exposures to chemicals, fumes, vapors, dusts, bloodborne pathogens, and ergonomic problems.

- Coordinating establishment of procedures for routine medical tests for employees pertaining to their working environment exposures.
- Reviewing adequacy of medical facilities and services such as testing equipment, first aid kits, and shift medical services.

TYPICAL ASSIGNMENTS

As with all other task groups, each member should have an assignment. Assignments that would be typical for Health and Environment Task Group members would include the following:

- *Ergonomics:* This individual would coordinate the ergonomics effort. It would also be appropriate for this member to chair the plant ergonomics team if there is one.
- *Spill/release prevention:* Employers should review systems, equipment, and procedures to identify vulnerabilities that might result in chemical spills and releases. This individual would examine these evaluations.
- *Hearing conservation:* This individual would review the hearing conservation program annually to ensure compliance and note any changes that might affect employee exposure to increased noise.
- *Hazard communication:* This individual would conduct an annual review of the company hazard communication program, comparing the written program to actual performance to assure continued compliance.
- *Respirators:* If respirators are used by workers, there must be a comprehensive respiratory protection program. This task group member would review the written program and respirator usage by employees along with maintenance, storage, and other functions to assure compliance.
- *Ventilation:* If there are ventilation systems, someone should be responsible for ensuring they are functioning properly. This means coordinating system functionality testing and testers to meet requirements.
- *Medical records:* Medical records can include medical and employment questionnaires or histories, including job descriptions and occupational exposures; preemployment or preplacement medical examination results; first aid records;

and employee medical complaints. If medical records are kept on site, this individual would be responsible for proper filing, maintenance, and security in the absence of a medical professional.

- *First aid:* This task group member would be responsible for the first aid program, ensuring that adequate first aid supplies are on hand and maintained and proper first aid records are maintained.
- *Permissible exposure limits:* What are the permissible exposure limits (PELs) for contaminants to which employees are exposed in the facility? This individual would be responsible for keeping up with permissible exposure limits and potential overexposures by monitoring permits and other documentation showing air test results. If the employer has an industrial hygienist, who would normally be responsible for this function, this task group member would work with the IH in some capacity to remain involved and assist with the management of PEL exposures.
- *Wellness:* If a wellness program is implemented, this individual would be the coordinator, possibly chairing a wellness committee or working with the plant nurse to assist in program effectiveness.
- *Personal protective equipment:* When personal protective equipment (PPE) is needed, it is necessary to ensure that the right type of PPE is selected and distributed. Hazard evaluations are required to help make such determinations. This task group member would coordinate the PPE program. In a large plant, it may be necessary to assemble a PPE committee. If so, this individual would chair the committee reporting back to the Health and Environment Task Group.

MEMBERSHIP

Because of the specialized nature of occupational health, industrial hygiene, and environmental protection, it is vital that some of the members of this task group have good working knowledge of environmental programs, occupational health programs, and industrial hygiene. A physician, plant nurse, or—in the absence of medical professionals—a first responder, should serve as an advisor to this task group.

MONTHLY REVIEWS

Each month when the Heath and Environment Task Group meets, they should be prepared to discuss the current status of a particular issue for which they are responsible. Task group members who have been given assignments should be prepared to lead the discussion during the month when their specific assignment comes up for discussion.

Figure 9-1 provides a list of suggested monthly review topics that may vary from plant to plant. Each topic listed is intended to be the major review item for the meeting that month. Topics can also serve as member assignments.

ERGONOMICS TEAM

The Ergonomics Team of the Health and Environment Task Group is responsible for the recognition, evaluation, and control of ergonomic hazards in the workplace. This team has special training enabling them to conduct the necessary reviews and evaluations of records and tasks to identify potential ergonomic stressors and make recommendations that will reduce or eliminate these problems. The team will

- Respond to employee concerns regarding suspected ergonomic stressors
- Conduct periodic surveys
- Conduct evaluations of new tasks and processes to identify any ergonomic problems or potential problems as early as possible
- Assist in training of employees in the fundamentals of ergonomics and the varied effects of poor ergonomics

Through the Health and Environment Task Group, the team will keep management abreast of ergonomic matters, enabling management to make informed decisions regarding any changes or purchases that may be needed.

The team chairperson is a member of the Health and Environment Task Group. This team chairperson, who is offen an engineer, should periodically accompany the Health and Environment Task Group chairperson to the FSHC meeting and provide an ergonomic update.

Task Group Monthly Review Items					Page 1 of 1
					Date:

This sheet is for use by task groups in establishing a review schedule to ensure that those items needing review are properly reviewed and reported to the FSHC.

Task Group: Health and Environment **Chairperson:** **Year:**

Month	Review Item	Reference	Assigned To	Comments/ Suggestions	Date Completed
January	Medical facilities				
February	Ergonomics				
March	Spill/release prevention				
April	Hearing conservation				
May	Hazard communication				
June	Respirators				
July	Ventilation				
August	Medical records				
September	First aid				
October	Permissible exposure limits				
November	Wellness				
December	Personal protective equipment				

Other Special Assignments:

Other Comments:

Figure 9-1. Task group monthly review items.

ASSIGNMENTS

The chairperson should give each team member individual assignments to coordinate. The assignment duration is a minimum of one year. As team members continue working on their assignments, they soon become very knowledgeable in that responsibility and a greater asset to the team and the organization. For that reason, ergonomic team members seldom rotate as often as do task group members. The following are examples of assignments for team members:

- Tools
- Seating
- Lighting
- Lifting
- Noise
- Employee feedback
- Workstation design/arrangement
- Repetitive motion
- Vibration
- Walking/working surfaces
- VDTs
- Personal protective equipment

As the need may dictate, other assignments may also be developed.

MEETINGS

The ergonomics team meets once each month. The meeting should last no longer than an hour. The group should spend this time reviewing progress and discussing newly recognized or potential problems and how they will be resolved. A list of monthly review items is provided at figure 9-2.

ENVIRONMENTAL TEAM

The Environmental Team of the Health and Environment Task Group is responsible for coordinating the facility's environmental protection program, as well as compliance with EPA regulations. The team has special training that enables them to recognize, evaluate, and control environmental hazards and to establish EPA compliance programs. The team will

- Evaluate potential environmental hazards throughout the workplace relating to air, water, and ground contamination

Ergonomics Team Monthly Review Items				Page 1 of 1 Date:	
This sheet is for use by Ergonomics Team in establishing a review schedule to ensure that those items needing review are properly reviewed and reported to the Health and Environment Task Group and FSHC.					
Task Group: Ergonomics Team Health and Environment		**Chairperson:**		**Year:**	
Month	**Review Item**	**Reference**	**Assigned To**	**Comments/ Suggestions**	**Date Completed**
January	Tools				
February	Seating				
March	Lighting				
April	Noise				
May	Lifting				
June	Workstation design/arrangement				
July	VDTs				
August	Repetitive motions				
September	Vibration				
October	Walking/working surfaces				
November	Employee feedback system				
December	Personal protective equipment				
Other Special Assignments:					
Other Comments:					

Figure 9-2. Ergonomics team monthly review items.

- Develop effective control measures for recognized potential environmental hazards
- Assist in developing compliance programs for applicable environmental regulations and regulatory compliance programs

- Routinely review and audit the effectiveness of the facility's environmental protection and regulatory compliance programs.

Through the Health and Environment Task Group the team will keep management abreast of environmental matters, enabling management to make informed decisions regarding environmental protection.

The team chairperson is a member of the Health and Environment Task Group and usually a person knowledgeable of environmental programs and regulations. He/she should periodically attend the Central Safety and Health Committee meetings with the Health and Environment Task Group chairperson and provide an environmental update.

ASSIGNMENTS

The chairperson should give each team member an individual assignment to coordinate. The assignment duration is a minimum of one year. As team members continue working on their assignments, they soon become very knowledgeable in that responsibility and a greater asset to the team and the organization. For that reason, environmental team members seldom rotate as often as task group members do. The following are examples of team member assignments.

- Air
  Clean Air Act (CAA)
  Indoor analysis
  Air analyses
  National Ambient Air Quality Standards
  Chlorofluorocarbons (CFCs)
  Fugitive emissions
  Toxic Substances Control Act
  Accidental release prevention
  Emission reduction
- Water
  Clean Water Act (CWA)
  Safe Drinking Water Act (SDWA)
  Federal Insecticide, Fungicide, and Rodenticide Act (FIFRA)

Environmental Team Monthly Review Items					Page 1 of 1 Date:

This sheet is for use by Environmental Team in establishing an annual review schedule to ensure that those items needing review are properly reviewed and reported to the FSHC.

Task Group: Environmental Team Health and Environment		Chairperson:		Year:	
Month	**Review Item**	**Reference**	**Assigned To**	**Comments/ Suggestions**	**Date Completed**
January	Clean Air Act (CAA)				
February	Clean Water Act (CWA)				
March	Safe Drinking Water Act (SDWA)				
April	Toxic Substances Control Act (TSCA)				
May	Federal Insecticide, et. al Act (FIFRA)				
June	Resource Conservation and Recovery Act (RCRA)				
July	CERCLAct (Superfund)				
August	Indoor air quality				
September	Air emissions				
October	Stormwater discharge				
November	Waste reductions				
December	Spill/Release Prevention Program				

Other Special Assignments:

Other Comments:

Figure 9-3. Environmental team monthly review topics.

Marine Protection, Research, and Sanctuaries Act (MPRSA)
Stormwater discharge
Release reduction
Spill prevention program

- Ground
  Resource Conservation and Recovery Act (RCRA)
  Comprehensive Environmental Response, Compensation, and Liability Act (CERCLA or Superfund)
  Waste reduction
  Used oil
  Solid-waste handling
  Spill prevention program
  Other (as needed)
- Training

Some team members will know little about environmental protection and their individual assignments. Thus, there will be a need for environmental training for team members. There are a number of sources of additional training and training material. See the list of environmental resources in Appendix C.

MONTHLY REVIEWS

A list of suggested monthly review items can be found in figure 9-3.

10

The Fire and Emergency Task Group

In January 2003, an explosion tore through a pharmaceutical supplies manufacturing plant in Kinston, North Carolina. Three employees were killed immediately, while three others died later from their injuries.

This was not an employer who neglected employee safety. To the contrary, this employer had many good components within its EHS system. Still, there were some things that simply fell through the cracks and were overlooked. And though the employer had a good safety–health system, initial OSHA fines exceeded $600,000.00. In a settlement agreement the company was fined $100,000.00 for failure to train employees and agreed to donate $300,000.00 to organizations who offered help at the time of and following the disaster.*

* Kinston Fire, West Settles Kinston Inquiry, July 17, 2003–May 24, 2004 (<http://www.newsobserver.com/news/kinstonfire/story/1260420p-7376921c.html>).

Effective Environmental, Health, and Safety Management Using the Team Approach, by Bill Taylor
Copyright © 2005 John Wiley & Sons, Inc.

This employer had frequent fire drills, but the training received by employees through drilling offered little help when suddenly they had to exit a facility surrounded by burning debris, twisted steel, and falling walls. This was something that had been neither practiced nor anticipated. The event would only have been made worse had it occurred at night when conditions would have made visibility difficult at best.

Every workplace is susceptible to emergencies. Through diligent adherence to codes and regulations, we can reduce the likelihood of such events, but in spite of best efforts, fire can break out at any time. Explosions can erupt with deadly consequences. Severe weather can endanger workers, along with chemical spills and releases or violent acts. Employers must prepare for disasters so that if an emergency does occur, employees will know the proper response. Employers should have written plans in place that leave no question about responsibilities and how employees will be safeguarded.

PURPOSE

The purpose of the Fire and Emergency Task Group is to ensure that plans are in place that will maximize employee and property protection from fire or explosion, chemical release, violence, or natural disaster.

PROGRAM IDEAS AND SUGGESTED ACTIVITIES

The Fire and Emergency Task Group must ensure that effective emergency and fire prevention and protection is established and maintained. Some activities to consider are

- Review existing or potential fire hazards and emergency situations.
- Analyze the adequacy of current fire protection equipment and procedures, including fire extinguishers, sprinkler systems, fire water supplies and distribution system, fire alarm systems, and fire walls, doors, and dampers.
- Establish fire control procedures for smoking, welding, burning, and any other open flame or ignition sources.

- Establish training, and supplying effective fire brigades or emergency teams.
- Routinely test emergency and fire protection equipment and procedures to ensure proper operation.
- Evaluate new equipment and facilities concerning potential fire hazards.
- Review facilities and operations for compliance with National Fire Protection Association codes, National Electrical Code, and Environmental Protection Agency regulations.

TYPICAL ASSIGNMENTS

Each member of the Fire and Emergency Task Group should be assigned specific responsibilities to review and coordinate. Once these assignments have been made, members should receive special training to ensure that they are knowledgeable about their assignments. Usually this training involves reviewing the applicable OSHA standard and NFPA codes. For example, a member assigned the task of reviewing emergency exits should receive training on OSHA standard 29 CFR 1910.33 through 37, and NFPA code 101—Life Safety Code.

Specific task group member assignments include responsibility for the following areas:

- Fire alarm system
- Emergency evacuation
- Emergency exits
- Fire extinguishers
- Sprinkler systems and hose lines
- Fire walls, doors, and dampers
- Flammable liquids
- Flammable gases
- Fire brigades and emergency teams

Members assigned these tasks should routinely review their assignments and report on their findings, conclusions, and recommendations to the task group. When questions or concerns arise involving items assigned to group members, those issues should be referred to the responsible group member for coordination. In this way the member's knowledge and expertise can be effectively utilized.

MEMBERSHIP

Although membership on this task group should be open to anyone with an interest, it is usually those employees and supervisors who serve as volunteer firefighters within the local community who enjoy the participation and bring good experience to the group.

Other effective members include those from maintenance, security, operations, or members of the facility emergency response team.

Normally the task group serves as an advisory board for the brigade or team and gives it direct, monthly contact with the line management organization.

MONTHLY REVIEWS

The Fire and Emergency Task Group meets each month to discuss the various elements of the fire–emergency program. Figure 10-1 is a suggested list of review items for which the task group is likely to be responsible. The list can be changed to fit the needs of the individual facility. Each item also would typically represent a task group member assignment that is to be covered in the meetings to ensure that issues are reviewed at least annually.

ADDITIONAL ASSISTANCE

Many employers in the United States have already established emergency response plans and procedures; however, many of these plans were created decades ago and have not been updated since. Or many plans simply were not adequate from the start and are in need of fine tuning. Following are suggestions offered to those who have no plans and must start from square one, or those who simply want to update existing plans.

FIRE PREVENTION PLAN

The best protection from fire or explosion is to prevent it altogether, so start with a fire prevention plan as required by OSHA. The plan should identify all potential fire source hazards. This is pretty straightforward. What is there in the facility that has a potential

Task Group Monthly Review Items					Page 1 of 1 Date:

This sheet is for use by task groups in establishing a review schedule to ensure that those items needing review are properly reviewed and reported to the FSHC.

Task Group: Fire and Emergency		Chairperson:			Year:
Month	**Review Item**	**Reference**	**Assigned To**	**Comments/ Suggestions**	**Date Completed**
January	Fire brigade/emergency team				
February	Medical				
March	Natural disasters				
April	Spills/releases				
May	Fire water systems				
June	Fire extinguishers				
July	Fire alarms				
August	Fire walls, doors, dampers				
September	Flammable/combustible liquids				
October	National fire prevention month activities				
November	Flammable gases				
December	Home fire safety				

Other Special Assignments:

Other Comments:

Figure 10-1. Task group monthly review items.

for igniting into fire or an explosion? Look for things such as welding, cutting, and burning operations; boilers and other sources of heat; smoking areas; exothermic compounds or processes; combustible storage or disposal areas; and electrical sources. These should be listed and controls for each identified as a part of the facility fire prevention plan.

Determine control methods for the accumulation of flammable and combustible waste material. Trash cans containing oily or solvent laden rags and solid waste should be emptied daily. Walls and surfaces should be kept clean of dust and overspray. Storage of waste should be kept to a minimum before disposal rather than allowing 55-gallon drums of hazardous waste to collect in those out-of-the-way places where they may be forgotten about for years.

There should be procedures established for regular maintenance of safeguards installed on heat producing and other equipment that can present a hazard. As many employers have strived to reduce costs, they have cut into the workforce (downsized), often reducing the number of maintenance personnel. Consequently, some of the routine, noncritical maintenance procedures have been moved down on the priority list simply because there are not enough maintenance workers to perform all the preventive maintenance and still do what is necessary to keep production machinery running. For this reason many things such as safety equipment, warnings, gauges not essential to operation, preventive maintenance on equipment, or drainage of compressor tanks and pressure vessels have gone unattended and fallen into disrepair. The disaster in Bhopal, India was a prime example of this.

Include as a part of the plan whom specifically, by name or job title, can be contacted with questions about equipment intended to prevent or control sources of ignition or fires and control of fuel source hazards. It is suggested that either job titles be used for this purpose or job titles and names. Names alone can make updating more difficult as people come and go more quickly than job titles change. Regardless, there should be a periodic review of the fire prevention plan, no less frequently than annually.

EMERGENCY ACTION PLAN

How will the workforce respond when there is an emergency? Will employees know what to do? Where to go? Whom to contact? This is all part of the emergency action plan.

Consider first the types of emergencies that can be anticipated. Certainly fire can occur anywhere, so that will definitely be one.

Most employers can also anticipate chemical spills or releases. Are you sure that no chemicals are used in your operation? If there is a copy machine that uses copy toner in the office, then there are chemicals used in your operation. Copy toner, in fact, is the only office product, when used as intended, for which OSHA would require a MSDS.

Other emergencies that could be anticipated may be explosions, bomb threats, tornados, heavy ice and snow, chemicals released from a neighboring facility or a train derailment, power failures, and the list goes on and on. Determine what could reasonably be anticipated, and then decide how you want employees to respond. If, for example, the facility is 100 miles from the nearest airport, it would be unreasonable to anticipate a plane crashing into or near the facility. On the other hand, if the facility is adjacent to the airport or at the end of a runway, such anticipation would hardly be unreasonable. Employers cannot plan for every conceivable eventuality but should plan within reason for those emergencies that could occur.

OSHA lists six elements as minimum elements of an emergency action plan; these should be incorporated into the facility plan:

1. What are the procedures for reporting fires and other emergencies? How will local authorities such as police and fire be notified? How will employees be notified? Are there horns, lights, whistles, and voice? Be sure to include methods that can be detected by those who may be hearing or visually impaired, or who may be working in areas of loud noise that may make hearing of an alarm difficult.
2. Establish procedures for evacuation. It should be decided how employees should evacuate and where they should evacuate to, keeping in mind that evacuation routes may be blocked by fire or explosion.
3. There may be a need for some employees to remain behind as others evacuate in order to perform certain critical functions. Plans should include procedures for such employees to follow so that they, too, can safely evacuate.
4. There should be a "muster location"—a place of refuge where employees can gather, enabling someone to get a head count. For the safety of those who may be required to enter the facility for search and rescue, it is vital to get an accurate accounting of the location of all employees and visitors.

For this reason the importance of remaining at muster locations until given further instructions regarding returning
into the facility or leaving the grounds must be stressed to
employees.

5. If employees will conduct rescue and/or medical operations,
there should also be procedures for them to follow.

6. Finally, who specifically can be contacted by employees to
get additional information or ask questions about the plan
or responsibilities under the plan?

Both plans should be reviewed periodically or whenever there are
changes in either facilities, operations, or personnel. Also determine how employees will be trained to be familiar with emergency
plans and fire prevention efforts. Do not assume that employees
are familiar with warnings and alarms. Hold drills as often as is
necessary to enhance employee understanding and proficiency in
carrying out responsibilities. Throw a wrench into the works
occasionally by simulating a blocked exit route or locked exit
door during a drill.

FIRE PREVENTION AND PROTECTION AUDITS

As they say, the best defense is a good offense. By conducting periodic audits of fire prevention and protection methods, employers
are less likely to be struck by fire, but if fire should occur, they will
be prepared.

To determine management's and employee's knowledge,
understanding, and implementation of fire prevention practices,
members of the Fire and Emergency Task Group should conduct
routine workplace audits. During these audits, the group members
should perform several tasks:

• Interview managers, supervisors, and employees.
• Observe facilities and work areas.
• Review emergency procedures.
• Check emergency response effectiveness.

Personnel conducting the audits should be trained and knowledgeable about the organization's emergency plans and procedures. Questions that they should ask individual managers,
supervisors, and employees include the following:

- What are the major fire hazards?
- How are the fire hazards controlled?
- Who checks to make sure that the controls are effective?
- Are emergency alarms known and understood?
- Are emergency reporting systems known?
- Are evacuation procedures posted?
- Are fire doors operable?
- Are emergency exits posted, operable, and unobstructed?
- How are medical emergencies handled?
- Are emergency lights provided and operable?
- Are fire ignition sources controlled?
- What would be done if a fire or other emergency situation were observed?

The answers given to these questions will provide needed insight to task group members regarding the adequacy and effectiveness of emergency equipment, procedures, and response. Depending on the answers, recommendations will be made to the head of the task group. Once the task group agrees on a set of recommendations, the head of the task group presents the recommendations at the next Central Safety and Health Committee meeting. Prior to presenting the recommendations, he/she should communicate the recommendations to the various departments audited.

11

The Housekeeping Task Group

Many fractures, sprains, and strains occur in the workplace because of slips, trips, and falls, which are usually the result of poor housekeeping. Poor housekeeping can also lead to fires, explosions, contamination, and blocked exits. Clearly, housekeeping is one of the vital elements of an effective employee EHS system and should be managed effectively.

PURPOSE

The purpose of the Housekeeping Task Group is to ensure that housekeeping standards have been established and are being practiced throughout the facility.

PROGRAM IDEAS AND SUGGESTED ACTIVITIES

The Housekeeping Task Group has very broad responsibilities in evaluating, coordinating, and making recommendations for controlling facilitywide housekeeping and orderliness. These broad responsibilities are to ensure that proper, uniform housekeeping and orderliness standards are established and maintained by all work areas and the entire operation. Some suggested activities for this task group are to

- Ensure that proper workplace housekeeping and orderliness standards are established and effectively communicated to all managers, supervisors and employees.
- Routinely audit the level of housekeeping and orderliness maintained by work area managers and supervisors and make appropriate suggestions for improvement. See the suggested housekeeping rating form in Figure A-11.
- Provide appropriate recognition systems for informing managers, supervisors, and employees of outstanding work area good housekeeping and orderliness.
- Ensure that management, supervisory, and employee education and training programs contain information and techniques for maintaining good workplace housekeeping and orderliness.
- Audit and review job practices and procedures to ensure that they include proper instruction for maintaining good housekeeping and orderliness.
- Provide publicity and communications systems for keeping managers, supervisors, and employees informed about the importance of management's concern for good housekeeping and orderliness.

THE HOUSEKEEPING TASK GROUP

The Housekeeping Task Group must conduct routine, uniform workplace housekeeping audits in order to ensure that consistent and proper evaluations and rankings can be made. Since monthly audits must be made by task group members, the group must have a sufficient number of members to minimize the audit time for any individual group member. In most cases two members team up to conduct an audit of a part of the facility. The number of teams depends on the number of facility areas that are to be audited each month.

To ensure that the team understands the desired level of housekeeping and orderliness to be maintained throughout the facility, special training must be given. Following a review of the information in this text, an on-site walkthrough orientation should be conducted.

The on-site walkthrough orientation should be attended by the entire task group. When possible, the ranking facility manager should accompany the team during part or all of the tour. The ranking manager's input is needed because the orientation tour is intended to make clear to each task group member the desirable level of housekeeping and orderliness. This tour will help ensure a uniform evaluation system for housekeeping throughout the facility. The task group members should record their evaluations. The entire group observes a situation and discusses the specific evaluation: a consensus level. Future visual observations of similar conditions should produce similar evaluations. This should promote standardized results among individual group members when they conduct housekeeping audits. Once the head of the task group concludes that all group members understand the evaluation system and will record similar evaluations, the walkthrough orientation can be terminated.

CONDUCTING HOUSEKEEPING AUDITS

The entire facility should be divided into audit areas, sufficient in number to permit a monthly audit of all operating areas. Some areas should not require monthly audits and may instead be audited quarterly or semiannually, depending on the nature of the operations. Some areas inside the facility fence but outside the building(s) may require only semiannual audits.

Normally two task group members are assigned to perform audits together. This enables the members to discuss their observations and to arrive at a consensus evaluation as they conduct the audit. An audit assignment sheet should be completed by the group chairperson and distributed to each group member.

Copies of the housekeeping audit sheet should be distributed at the Facility Safety and Health Committee meeting by the Housekeeping Task Group chairperson. Both the housekeeping monthly audit system and the sheet should be discussed at this meeting.

All departments should be informed of the first two monthly audits and given an opportunity to accompany the auditors in order to learn how the audit is being conducted and to observe the evaluation system. This will allow each department to conduct its own housekeeping audits if desired.

After the first two audits, all future housekeeping audits should be unannounced. Only the housekeeping auditors should know the specific audit day and time.

A uniform audit time should be established for all audits to ensure that evaluations are as equal (fair or impartial) as possible. The nature of the operations should also be considered; however, housekeeping defects must not create unsafe conditions under any circumstances.

Copies of housekeeping audit reports must be submitted to the area audited and to the task group chairperson. All reports must be received by the chairperson by the first day of each month so that a summary report can be prepared in time for the next Facility Safety and Health Committee meeting.

REPORTS

Each area of the facility audited is assigned a specific numerical value by the auditors. The summary report prepared by the head of the task group should rank the areas audited, with the best area appearing at the top of the list.

A copy of the summary report should be given to each Facility Safety and Health Committee member and to the committee secretary for filing with the meeting minutes. Consideration should also be given to the idea of posting the summary report on safety bulletin boards throughout the facility.

Repeat defects recorded on the auditors' reports should be highlighted on the summary report. When the same defect is recorded for three consecutive months, a special housekeeping evaluation should be submitted with the manager responsible for the area, and recommendations for correcting the problem should be made.

ASSISTANCE

All facility areas should be encouraged to solicit help from the Housekeeping Task Group in solving workplace housekeeping problems. A "request for assistance" form is shown in Figure A-12.

RECOGNITION

The Housekeeping Task Group should officially recognize areas of the facility that exhibit outstanding housekeeping and orderliness. This is accomplished on a monthly basis by the ranked list of audited areas in the task group chairperson's summary report to the Facility Safety and Health Committee. However, additional recognition may be provided quarterly or annually for continuing excellence. These awards may take the form of a banner or plaque that can be displayed in the area. An annual award may be presented by the ranking manager at a special luncheon or dinner. Housekeeping Task Groups wanting to implement such recognition are encouraged to work with the Activities Task Group to develop an appropriate recognition award or method.

PROCEDURES

Housekeeping practices must be included in all job procedures. To that end, the Housekeeping Task Group should routinely review job procedures to determine the adequacy of their housekeeping and orderliness provisions. All personnel responsible for preparing job procedures should be made aware of the need for housekeeping instructions.

PUBLICITY

Organization magazines, newspapers, and bulletins should be used to publicize the importance of good housekeeping to the organization and to employees. Photos showing outstanding housekeeping should be used when possible to illustrate good performance.

TRAINING

Training programs for new and current employees should include coverage of the expected level of housekeeping and the methods to be used in achieving and maintaining the expected level. Employees must understand that workplace housekeeping and orderliness

Task Group Monthly Review Items				Page 1 of 1 Date:	
This sheet is for use by task groups in establishing a review schedule to ensure that those items needing review are properly reviewed and reported to the FSHC.					
Task Group: Housekeeping		Chairperson:		Year:	
Month	Review Item	Reference	Assigned To	Comments/ Suggestions	Date Completed
January	Annual audit schedule				
February	Operating areas				
March	Offices				
April	Warehouses				
May	Outside/yards				
June	Storage cabinets				
July	Vehicles				
August	Remote areas				
September	Overhead areas				
October	Shops				
November	Electrical rooms				
December	Parking lots				
Other Special Assignments:					
Other Comments:					

Figure 11-1. Task group monthly review items.

are a personal responsibility. Housekeeping is a part of every job and must be done on a continuing basis. Housecleaning is not housekeeping. If good housekeeping is performed, housecleaning is not necessary.

TYPICAL ASSIGNMENTS

Unlike other task groups, the Housekeeping Task Group routinely goes through the workplace in order to conduct an evaluation. Their function is different from, yet no less important than, that of other task groups, and their assignments are different. Typical assignments for Housekeeping Task Group members will be the locations that they will be assigned to inspect each month. The Task Group Review Items checksheet shown in Figure 11-1 is a suggested meeting format for monthly meetings topic listing areas within the facility that should be inspected, but these same areas may serve as area assignments for task group members. Reviewing the items on this list will help ensure that each area is discussed in terms of housekeeping issues.

Some users of the CTJ Facility Safety and Health system will have task group members conduct their monthly inspections and then meet to discuss findings or a particular location. Others, however, where time is critical, may instead let the monthly inspection tour suffice for the monthly meeting. It may still be necessary for the task group to meet briefly from time to time, but either way can be effective as long as inspections are effective.

MEMBERSHIP

The Housekeeping Task Group works better when employees from every area are represented. This not only enhances the quality of inspections but also helps with communication.

Members should also be made aware that it is not their mission to correct housekeeping problems. Problems should be brought to management's attention through the Facility Safety and Health Committee, but task group members should not take corrective actions upon themselves.

The Incident Investigations
Task Group

Hazards are typically identified in one of two ways: (1) through an inspection or (2) after an incident or near miss has occurred and an investigation is conducted. The primary reason for conducting investigations is to prevent a recurrence of the incident. For this reason the incident investigation process is very important and should be effectively managed.

PURPOSE

The purpose of the Investigations Task Group is to monitor the quality of investigations of incidents and near misses, ensuring that all causes have been determined and appropriate corrective action taken.

Effective Environmental, Health, and Safety Management Using the Team Approach, by Bill Taylor
Copyright © 2005 John Wiley & Sons, Inc.

PROGRAM IDEAS AND SUGGESTED ACTIVITIES

Some activities that the Investigations Task Group may get involved with are

- Routinely review and properly maintain effective accident investigation procedures.
- Ensure that all line supervisors are properly trained to conduct effective incident investigations.
- Review all incidents, maintain statistics, and analyze findings to determine needs.
- Recommend needed improvements to the Facility Safety and Health Committee.
- Assign task group members to participate in special incident investigations and report findings to the task group.
- Once the basic causes of incidents have been determined, the incident investigation findings should be referred to other task groups for follow-up (e.g., safety rule violations to the Rules and Procedures Task Group).
- Obtain pertinent incident data and specific incident descriptions from other similar facilities and communicate throughout the facility.
- Routinely publicize serious incidents with serious potential throughout the facility.
- Request and coordinate follow-up action by all facility groups, where applicable.

OPERATING PROCEDURE

The head of the task group arranges to have copies of all injury and noninjury incident investigations sent to the task group for review. Once received, the task group head screens the reports and selects specific incidents that exhibit serious potential for follow-up. These reports are then assigned to task group members for follow-up. They review the report and discuss their conclusions and recommendations with supervisors and employees in the work area where the incident occurred.

Results of the task group review are shared with the head of the department where the incident occurred and are reported to the Facility Safety and Health Committee.

TYPICAL ASSIGNMENTS

Assignments given to task group members to ensure high-quality incident investigations include the following:

- Incident investigation procedures
- Incident records and statistics
- Noninjury incidents
- Incident investigation training
- Incident trends

MONTHLY REVIEWS

The Investigation Task Group should review certain factors each month in an effort to ensure that the investigation process is still functioning as it should. A suggested list of review items can be found in figure 12–1.

INVESTIGATION REPORT REVIEWS

The typical incident investigation report is broken into sections. The Incident Investigations Task Group should review each report by section to ensure that they are properly completed. This review will help to determine whether all causes and contributing factors have been identified and if suggested corrective actions are adequate.

BASIC INFORMATION

The basic information includes things such as date of the incident; name, age, and gender of the injured person or employee involved; location of the incident; and time of occurrence. All of this information is important and should not be omitted.

DESCRIPTION OF THE INCIDENT

The description provided in the report should be sufficient to allow the reviewer to have a clear picture of what occurred. Something went wrong, or there would have been no incident. The description should clearly but concisely provide this information, identifying what exactly went wrong.

Task Group Monthly Review Items					Page 1 of 1 Date:
This sheet is for use by task groups in establishing a review schedule to ensure that those items needing review are properly reviewed and reported to the FSHC.					
Task Group: Incident Investigation		**Chairperson:**		**Year:**	
Month	Review Item	Reference	Assigned To	Comments/ Suggestions	Date Completed
January	Annual incident summary				
February	No-injury accidents				
March	Report promptness				
April	Witness statements				
May	Photos/sketches/etc.				
June	Engineering controls				
July	Education/training				
August	Enforcement				
September	Basic causes				
October	Recommendations				
November	Management reviews				
December	Off-the-job incidents				
Other Special Assignments:					
Other Comments:					

Figure 12-1. Task group monthly review items.

CAUSE(S)

Rarely will there be a single cause to an incident. If the report lists only one cause, then it probably is not complete. The report should identify a cause or causes but should also include any contributing factors. Something as simple as an employee running up the stairs who falls and fractures an arm can have several contributing factors. Why was the employee running? Was he/she in a hurry? If so, why? What was going through this person's mind? Was he/she thinking about why he/she was running? These are all questions that could yield a fertile field of cause and contributing factors. A thorough investigation will explore these and other possibilities.

CORRECTIVE ACTION

Once the cause(s) has (have) been identified, it is time to get it (them) fixed. The report should list any corrective actions suggested that will prevent a recurrence. Often an investigation will uncover issues that did not cause or contribute to the incident under investigation, yet it is something that could result in future incidents if not corrected. These, too, must be listed and corrective actions listed with them.

It is important to closely scrutinize this section because an investigator, in a hurry to complete an investigation, seldom will give adequate thought to the best corrective actions, even though it might be a simple fix.

A good example is an employee who stepped in a low spot (depression) in the sidewalk that had filled in with mud following a recent rain. The employee slipped and injured his knee. The problem was properly identified by the investigator, but as for corrective action, the investigator simply in wrote that the supervisor should instruct his employees to be more careful of low spots in sidewalks.

With just a little more thought, the investigator may have determined that future incidents of this type could better be prevented by eliminating the depression. Fill in the low spot with a filler material such as concrete or concrete patching compound, thereby eliminating the hazard.

REVIEWS

Each completed report should be reviewed by the department head, EHS professional, and ranking manager at the facility. Each should sign and date the report. During the task group review the dates should be compared to the date of the investigation. Reports should be reviewed and signed off within a week following the investigation if possible.

At the same time, compare the investigation date with the date of the incident. Investigations should begin as soon after the incident as possible. If more than a day or two has gone by before the investigation begins, then it is taking too long and corrective action is needed.

EVALUATING OCCUPATIONAL HAZARDS

Effective evaluation of occupational hazards requires careful consideration and in many cases special knowledge and training. Proper evaluation of situations takes into consideration the type of hazard, injury and illness potential, frequency of exposure, and experience. One major shortcoming of an ineffective evaluation is to assume that an operation, material, or practice is safe simply because injuries and illnesses have not occurred in the past. Experience has proved over the years to be a poor predictor; therefore, an employer should not rely on experience when attempting to evaluate the safety of an operation, task, or piece of equipment.

Today, special industrial hygiene instrumentation and laboratory analyses are required to properly evaluate many occupational health hazards. Companies must develop the required in-house expertise or hire outside consultants to effectively evaluate job health hazards. All chemical substances must be considered potentially toxic under certain conditions and proper evaluations conducted.

CONTROLLING OCCUPATIONAL HAZARDS

Once management has recognized and effectively evaluated occupational hazards, proper controls must be implemented. Occupational safety and health hazard controls usually involve the following:

- Engineering controls to eliminate the exposure through isolation, separation, ventilation, and guarding is the first and most important technique to consider. Where feasible engineering controls can be provided, they should always take precedence over other control methods.

- Education and training controls must always be utilized in conjunction with feasible engineering controls since employees can be injured or become ill regardless of the thoroughness of engineering controls. However, the combination of engineering and education controls can and usually do achieve a safe and healthful working environment.

- Enforcement, disciplinary procedures, and positive motivational programs must always be an important part of every EHS system. All personnel must be positively motivated to properly protect themselves by utilizing the engineering controls provided, and following the specified safety rules and procedures. Where employees fail to follow procedures, they must know that management will take appropriate corrective action. Failure of management to act promptly and effectively will defeat the entire EHS system.

13

The Security Task Group

Toward the end of the twentieth century the economy took a downward spiral, leading many employers to make significant cuts to reduce costs. Many of these cuts occurred in plant security as security forces were either reduced or curtailed completely.

Today workplace violence is perennially among the top three causes of death on the job in the United States, and number one among female workers. Industrial and economic espionage is a growing concern that costs American businesses billions of dollars in losses each year. While the majority of assaults are a result of domestic violence or disgruntled employees, terrorism is a growing concern and no longer considered to be a concern just for overseas workers. Terrorism and defending against it is a genuine threat to American workers that has caught our attention and encroached on the daily lives of many. Still, many employers are without a security presence.

There are many things employers can do toward meeting their own security needs. However, just like safety and health, it takes involvement of the workforce to make it work best.

PURPOSE

The purpose of the Security Task Group is to help increase employee awareness of undesirable events and supplement existing security objectives.

AWARENESS

Generally speaking, the larger the security force and more sophisticated the security efforts, the greater the secrets, service, or product the company is trying to protect. All employers do not need closed-circuit monitoring and foot patrols. But with any employer, there is a need for early identification of unwanted intruders or fire detection. There is a need to protect sensitive information and trade secrets and prevent pilferage. And there is most definitely a need to protect workers from acts of violence. But regardless of the magnitude of the security efforts, the best security can do only so much. If an intruder is very determined to get into the facility, there is always a way. If the right employee is offered the right temptation he/she will steal from the company, whether its trade secrets or a laptop computer.

Employees can fill a big role in the prevention of undesirable events. The presence of dedicated employees provides additional eyes and ears. It is not the responsibility of the Security Task Group to manage security where there exists a security force, or attempt to apprehend violators. Rather, the task group will simply coordinate efforts to increase employee awareness.

PROGRAM IDEAS AND SUGGESTED ACTIVITIES

Protecting facilities, employees, trade secrets, and so on, can be a daunting task. Every employee can and should be an extension of the security force by noticing and reporting unusual happenings or unidentified persons.

SECURITY NEEDS AND VULNERABILITY ASSESSMENT

Some facilities are well staffed and use state-of-the-art surveillance and monitoring equipment to detect intruders and protect against theft or vandalism. Most plants; however, have little or no security measures in place. Employers should conduct security assessment needs to determine what efforts should be taken to enhance security or what existing efforts should be improved. A security needs assessment can include the following:

- Does the company have trade secrets or information to protect?
- What methods are employed to protect them?
- Who is responsible for the protection of trade secrets?
- Does the employer provide a service or consumable goods to a large number of customers, such as water filtration and food and beverage manufacture?
- What safeguards are in place to protect the safety of the end user of the product or service?
- Are employees required to leave or be outside the facility for any reason such as taking measurements or samples?
- How are they protected from assault by intruders?
- Is lighting in the parking lot and around the facility adequate?
- Is fencing adequate and in good condition?
- Is additional protection (e.g., barbed wire or razor ribbon) needed?
- Are there vehicles or pieces of equipment parked within the fence and unguarded?
- Is visibility near exits clear and unobstructed?
- Is the roof accessible?
- Are doors and windows secure from outside intruders?
- Is it possible for individuals to enter the facility or grounds undetected or unchallenged?
- Are closed-circuit cameras or other electronic surveillance equipment used?
- Is surveillance equipment properly maintained?
- Is surveillance equipment monitored?
- Are there large quantities of flammable, combustible, or toxic gases or liquids stored on plant grounds and unprotected from vehicular traffic or unauthorized tampering?
- Are there large quantities of hazardous substances within close proximity of the neighboring community?
- Are employees instructed in techniques of proper disposal of sensitive information?

- Is trash compacted or shredded, or is it easily accessible to unauthorized persons?
- Is computer information safeguarded?
- Are visitors required to be escorted at all times when entering the facility?
- Does the company have an effective employee assistance program?
- Are contractors, including service contractors such as landscapers, janitors, and computer repair technicians, monitored?
- Do nested contractors conduct background checks or other employee "fitness for duty" assessments on employees who will enter the facility?
- Are locks checked periodically to ensure that gates and doors are not accessible to unauthorized personnel?
- Who is responsible?

SUGGESTED ACTIVITIES

Suggested activities include:

- Identify and review potential vulnerabilities.
- Review adequacy of current protection and surveillance systems and equipment.
- Review training and assess the needs of employees regarding security awareness.
- Evaluate procedures for handling and disposal of sensitive material and information.
- Routinely test surveillance equipment.
- Evaluate new facilities for employee protection from assault and security breaches.

RECOGNITION

Employees often feel that safety, health, environmental protection, and security are other persons' responsibilities and therefore tune out issues that they feel are not directly related to their own jobs and responsibilities. But because a lack of security could result in loss of business, loss of property, or loss of life, all employees should be instructed to be aware of and report anything that is different. This could be unauthorized individuals in the workplace, unattended packages, something missing, unusual odors or sounds, or other anomalies.

TYPICAL ASSIGNMENTS

Typical assignments include monitoring of the following:

- *Surveillance equipment*—check existing surveillance equipment to determine adequacy and function.
- *Facility accessibility*—is the roof accessible? Are windows, doors, and gates that are supposed to be locked secured? If not, why?
- *Access authorization*—is access to unauthorized persons controlled? Does the system function?
- *Waste disposal*—is sensitive documentation properly disposed of by shredding or burning?
- *Visitors and contractors*—do contractors check backgrounds of employees? Check to be sure that service personnel are monitored.
- *Lighting and landscaping*—does landscaping present good places where intruders or assailants can hide? Is lighting adequate to deter intruders or make them easily visible?
- *Trade secrets*—are there trade secrets that must be protected? How is this done?
- *Parking lots and fences*—is the parking lot safe for employees arriving or leaving after dark? Is the fence damaged, or does it have openings at the bottom created by animals or erosion?

MEMBERSHIP

Task group members should come from different areas, including maintenance, engineering, and human resources. Prior military and law enforcement employees usually have an interest and bring valuable experience to the Security Task Group. If the facility has a security force already, the leader or designee should serve as an advisor to the task group and the FSHC.

Task Group Monthly Review Items					Page 1 of 1 Date:
This sheet is for use by task groups in establishing a review schedule to ensure that those items needing review are properly reviewed and reported to the FSHC.					
Task Group: Security		Chairperson:			Year:
Month	Review Item	Reference	Assigned To	Comments/ Suggestions	Date Completed
January	Security alarms				
February	Facility accessibility				
March	Lighting and landscaping				
April	Trade secrets				
May	Vulnerability assessment				
June	Access authorization				
July	Education/training				
August	Visitors and contractors				
September	Parking lots and fences				
October	Vehicle passes				
November	Roof and windows				
December	Surveillance equipment				
Other Special Assignments:					
Other Comments:					

Figure 13-1. Task group monthly review items.

MONTHLY REVIEWS

Figure 13–1 contains a suggested list of items that should be reviewed at monthly meetings. The list is only a suggestion and may be changed or adapted to fit the needs of the organization. Each topic on the list also may serve as member assignments.

14

Safety and Health Responsibilities

There was a time when there was no such thing as OSHA. There were no laws requiring employers to guard machines or train workers in the ways of working safely, and while many employers had already realized the value, not to mention the moral and ethical responsibility to protect their workers, most failed to act in this regard. Thus, many thousands of workers were injured or killed on the job every year. Along came OSHA in 1970, and suddenly there was a federal law requiring employers to provide a safe and healthful workplace. Employers thought for the most part that this would require no more than fire extinguisher maintenance and an occasional training class in general safety. Many employers felt that the solution was to assign employee safety to the human resources manager (except back then they were called "personnel managers") as a collateral duty, or to turn the responsibility over to an individual who may be nearing retirement. For years, this mentality pervaded throughout industrial America.

So, traditionally, as the years have advanced, we have managed employee safety and health either through a committee that got together once each month to discuss the latest inspection findings, or by entrusting the entire system to an individual who may or may not have any managerial experience, formal training, or education in safety. These individuals, dedicated to the preservation of human life and protection of their fellow workers, were out there, often all alone with little more than oral support (lip service) from senior management. They could often be seen single-handedly trying to enforce laws handed down by Congress and seen by employers and employees alike as a major inconvenience at best, and would often find themselves as the most unpopular workers at the plant.

Employees did not want to wear safety shoes or hearing protection; they did not want to turn off the power and lock out the machine just to clear a jam. And unless management agreed that this policy was important and enforced it, the employee simply did not comply it. Employees did what they thought management expected of them, which usually, in the name of production, was contrary to safety rules. The "safety guy" had very little, if any, authority. As a result, safety programs at most companies existed in name only, and we had all these safety cops running around the facilities reminding employees on a daily basis to follow the rules. Safety and health got a bad name.

Times have changed. Managers are quickly becoming aware of the impact that safety and health have on the bottom line. They are beginning to realize the value in having good safety and health systems. But at the same time it is becoming obvious that an employee safety–health system is more than fire extinguisher inspections and annual training. They are becoming aware of what is required for an effective system that will protect workers along with the fact that it must be managed in the same way that everything else is managed.

Yes, there are still many safety and health managers out there with collateral responsibilities. In a small to medium-size company, that can work just fine, but the one responsible for managing workplace safety and health must delegate responsibilities, just as other responsibilities are delegated. When an employer assigns employees to conduct inspections, lead incident investigations, audit programs, write rules, and all the other things traditionally performed by safety and health professionals, the role of the professional changes drastically. Instead of spending time attending to the day-to-day needs of running the EHS system, checking fire extinguishers, or looking for lockout violations, they find themselves serving more as a consultant or advisor to employees on the front line.

That is the major responsibility of the EHS staff under the FSHC system. They should provide advice and guidance to task groups, task group members, and the Facility Safety and Health Committee.

For instance, an employee serving on the Inspections and Audits Task Group who has been assigned forklifts is not likely to know where to go for detailed information regarding inspection requirements for these vehicles. It would be up to the EHS staff to provide this information.

Additionally, staff members should keep management apprised of new and ever-changing safety, health, and environmental issues. If OSHA issues a new standard, for example, the staff EHS representative must become familiar with it and present an explanation to the FSHC so that responsibilities can be properly delegated.

Many issues will be common to multiple task groups. For that reason, the boundaries are not always clear and it becomes very easy for a task group to find itself duplicating the effort of another task group. The EHS representative, being a member of every task group, should be aware of what each group is doing and discussing in their monthly meetings. This will afford him/her the opportunity to oversee jurisdictional infringement.

The EHS staff representative serves as the secretary for the FSHC. This individual does not chair this committee, even in an acting capacity during the ranking manager's absence. Simply put, the EHS professional becomes the coordinator for the entire FSHC system, in much the same way that the "production manager" coordinates the production efforts or the "human resources manager" manages human resources.

The EHS professional is a key individual to the success and viability of this system. As the role takes on a different direction, it will become easier and more rewarding. Management at every level is responsible for ensuring that employees, supervisors, and task group members are held accountable for individual responsibilities. This includes compliance with rules and procedures as well as task group member assignments.

Top management may have every intention of having a top-shelf EHS system to protect workers and the environment. Unfortunately, for several reasons, by the time it gets down to the supervisor level, rules and procedures are seldom enforced. Because of the relationship between employees and their supervisors, or perhaps because of quotas or production demands, rules are simply ignored by employees and their failure to comply is likewise ignored by the supervisor.

It sometimes becomes a balancing act for the employee, the supervisor, or the department manager when it comes to meeting production demands and at the same time complying with the safety and health rules. And while production is certainly important, it should not be achieved or pursued at risk to the worker. A balance must be struck to ensure that the worker is protected. This is the responsibility of managers and supervisors. This is how world-class safety cultures are established.

An employer's incident rate is not a true measure of that employer's safety and health efforts. Too many things can happen that are beyond the employer's control that can elevate the number of injuries and thus the incidence rate. A neighboring employer can have a chemical release that drifts into the plant, sending several employees to the hospital following chemical exposure. A train can derail, releasing toxic substances or causing a fire, which can injure or kill employees inside the plant.

An employee sitting in the medical department or lunch room wearing a body cast while counting widgets is not an indicator of injury prevention. It may keep that injury out of the lost-workday column, but someone still got hurt. Rather than hang their hats on incidence rates, managers should be monitoring the safety–health atmosphere. What do employees commitment (including managers) think of safety? What do they do safety to demonstrate? How are they involved in the safety–health effort?

Safety–health is a responsibility of every individual at the facility regardless of what their job might be. This is one point that should be made clear to everyone. Managers and supervisors are as responsible for the safety of employees under their charge as they are for getting a quality product out the door. This, too, should be made clear. But none of this really matters if the ranking manager doesn't establish and enforce the mandate. If the incidence rate is being used as an indicator and the numbers are high, then management should first evaluate their own actions toward managing safety and then look right down the line. When a worker violates a rule, resulting in injury, managers should ask why the rule was violated. Management should ask why noncompliance is permitted.

A great deal of responsibility goes with the job of managing, and protecting employees is of no less importance than any other responsibility. When everyone realizes their responsibilities and EHS is managed in the same way that everything else is managed, employees are safer and injuries are fewer. You can take that to the bank.

Appendix A
Tools and Forms

This appendix contains a series of sample forms and other documents (Figures A-1 through A-12), all of which are discussed in the main text.

ANNUAL MEETING SCHEDULE
FACILITY SAFETY AND HEALTH COMMITTEE TASK GROUP

Task group												
Year												

Meeting dates												
Jan	Feb	March	April	May	June	July	Aug	Sept	Oct	Nov	Dec	

Meeting time												
Jan	Feb	March	April	May	June	July	Aug	Sept	Oct	Nov	Dec	

Meeting location												
Jan	Feb	March	April	May	June	July	Aug	Sept	Oct	Nov	Dec	

Figure A-1. Sample form for FSHC task group scheduling.

SAFETY AND HEALTH TASK GROUP MEETING MINUTES		Date:	
		Number:	
Task group:		**Chairperson:**	
Meeting date:		**Meeting location:**	
Meeting time:	**Began**	am/pm	**Ended**	am/pm

Attendees:

Absent:

Visitors:

Subject(s) discussed:

Recommendations:

Other items:

Next meeting		
Date:	**Time:**	**Location:**

Assignments:

Minutes prepared by:		**Date:**

Route copies to:	1—Task group members	2—Safety committee chairperson
	3—Safety coordinator	4—File

Figure A-2. Sample form for recording FSHC task group meeting minutes.

FSHC TASK GROUP MEMBER ANNUAL ASSIGNMENT	Date:

This sheet is provided to inform Facility Safety and Health Committee (FSHC) task group members of their annual assignments and specific details for completing assignments. For more information, contact the task group chairperson, safety office, or other resources.

Task group:	Organization:	Location:

Annual assignment:	Annual review month:

Member assigned:	Title:	Job:

General information

Each FSHC task group member is assigned an annual task to review, coordinate, and report on at task group meetings. The entire task group will review the annual assignment at least once during each year. At that time the task group member assigned the annual review responsibility will lead the entire task group review at that monthly task group meeting. To do this effectively, the entire task group should be reminded of the upcoming review at the preceding group meeting. Thus, each member will have an opportunity to review the assigned review item during the previous month.

During routine monthly task group meetings, the member should report on any pertinent items pertaining to his/her annual assignment. Also, any questions or problems occurring during the year concerning the member's assignment will be referred to the member.

The task group chairperson may request the member coordinating an annual review to accompany him/her to the next FSHC meeting and report on the results of the annual review.

Planning reviews

To ensure that the necessary basic information is obtained concerning each assignment, the member should discuss his/her assignment with the task group chairperson and safety/health coordinator. Some of the information that may be useful are as follows.

References

Safety Manual	ANSI
OSHA	Other
National Safety Council	Other

Monthly reviews

The member with the annual task group review assignment should keep alert to observe workplace situations throughout the month while he/she performs routine job assignments. Usually there are many opportunities to see or check on special assignments without taking time away from normal work assignments. Make notes of observations and discussions and report on findings, conclusions, and recommendations at monthly task group meetings.

Figure A-3. Sample form for task group member assignments.

FSHC TASK GROUP MEMBER ANNUAL ASSIGNMENT	Date:

This sheet is provided to inform Facility Safety and Health Committee (FSHC) task group members of their annual assignments and specific details for completing assignments. For more information, contact the task group chairperson, safety office, or other resources.

Task group: Inspections	**Organization:** CTJ Safety	**Location:** NC Office

Annual assignment: Forklifts	**Annual review month:** April

Member assigned: John Doe	**Title:** QA technician	**Job:** Test samples

General information:

Each FSHC task group member is assigned an annual task to review, coordinate, and report on at task group meetings. The entire task group will review the annual assignment at least once during each year. At that time the task group member assigned the annual review responsibility will lead the entire task group review at that monthly task group meeting. To do this effectively, the entire task group should be reminded of the upcoming review at the preceding group meeting. Thus, each member will have an opportunity to review the assigned review item during the previous month.

During routine monthly task group meetings, the member should report on any pertinent items pertaining to his/her annual assignment. Also, any questions or problems occurring during the year concerning the member's assignment will be referred to the member.

The task group chairperson may request the member coordinating an annual review to accompany him/her to the next FSHC meeting and report on the results of the annual review.

Planning reviews

To ensure that the necessary basic information is obtained concerning each assignment, the member should discuss his/her assignment with the task group chairperson and safety/health coordinator. Some of the information that may be useful is as follows.

References

Safety Manual—Section 7	ANSI
OSHA 1910.178(q)	Other: ASME B56.1-2000
National Safety Council	Other

Monthly reviews

The member with the annual task group review assignment should keep alert to observe workplace situations throughout the month while he/she performs routine job assignments. Usually there are many opportunities to see or check on special assignments without taking time away from normal work assignments. Make notes of observations and discussions and report on findings, conclusions, and recommendations at monthly task group meetings.

Figure A-4. Completed sample.

FSHC TASK GROUP REVIEW	Date:
	Page 1 of 2

This Facility Safety and Health Committee (FSHC) task group review form is designed for use by the chairperson of the FSHC or his/her designee to review the overall performance of the task groups. It may be performed annually or more often as needed to evaluate the progress or lack of progress and to recommend improvements when needed.

Organization: **Facility:**   **Location:**

Task group reviewed: **Period covered:**

| | From: | To: |

Task group chairperson: **Title:**

Department: **Service as chairperson:**

| | From: | To: |

Task group members

Name	Department	Assignment Date	Special Assignments/Comments/etc.

Review	Jan	Feb	Mar	April	May	Jun	July	Aug	Sept	Oct	Nov	Dec
Meeting dates												
Number of members attending												
Annual review item												
Recommendations made to FSHC												
Recommendations adopted by FSHC												
Special accomplishments												

Chairperson effectiveness

❑ Annual meeting schedule provided all members.

❑ Agenda prepared for each meeting.

❑ Discussed agenda with staff coordinator prior to monthly meeting.

❑ Attended each monthly task group meeting.

❑ Annual emphasis items reviewed monthly.

❑ All members assigned a specific item to coordinate and report on.

❑ Member reports requested at each meeting.

❑ Meeting minutes kept and distributed to each member and FSHC secretary.

❑ Number of FSHC meetings missed (   ).

❑ Priorities set for task group projects.

❑ Task group goals/objectives meeting.

❑ Other task group members attend at least one FSHC meeting.

Figure A-5. Sample task group review form. *(continued)*

	FSHC TASK GROUP REVIEW	Date:
		Page 2 of 2

Chairperson effectiveness (continued)

❑ Number of monthly task group meetings missed (   ).

❑ Meetings started and ended on time.

❑ Member training provided as needed.

❑ Reports presented at each CSHC meeting.

❑ Good meeting location provided.

❑ Other:

Member effectiveness

❑ Routine meeting attendance.

❑ Annual assignments handled effectively.

❑ Participates in discussions and recommendations.

❑ Attends at least one FSHC meetings.

❑ Makes effective reports to FSHC when asked.

❑ Discusses assignments with other personnel to get their views.

❑ Makes positive recommendations.

❑ Other:

Task group effectiveness

❑ Achieved goals/objectives.

❑ Contributed to safety and health process success.

❑ Cooperated with other task groups.

❑ Made positive recommendations that were usually adopted.

❑ Member safety and health awareness enhanced.

❑ Annual review items completed on schedule.

❑ Other:

Task group strong points

Task group items needing improvement

Recommended program for improvement

Review completed by:	Date:	Review approved by:	Date:
Review discussed with task group chairman by:		Date: ❑ Positive	Chairperson's reaction: ❑ Negative

Figure A-5. *(continued)*

	SAFETY MANUAL CHECKLIST	Date:
		Page 1 of 7
Employer:	**Facility:**	**Unit:**
Address:		

General information:

Every employer should have a safety and health policies and procedures manual. A written manual is important to

- Document company/governmental safety and health policies.

- Train and educate employees as to safety and health policies.

- Establish a safety–health management program that includes both management and hourly employees.

- Enforce safety policies fairly and consistently.

- Organize and address all safety and health issues.

This manual should be developed with input from all levels of the organization. Every policy should be written and feasible, employees trained, and the policies enforced. A number of resources are available to obtain prepared policies that can be adapted to a worksite. Those resources include

- Governmental regulatory agencies

- Industry associations

- Safety and industrial hygiene associations

- National Safety Council's Accident Prevention Manual

- OSHA, ANSI, NFPA, and other standards

- CTJ Safety Associates' OSHA Compliance Checklists

Figure A-6. Sample safety–health manual checklist. *(continued)*

Check When Complete		Subject	To Be Developed By	Target Date	Date Completed
		SAFETY MANUAL CHECKLIST Date: Page 2 of 7			
	I.	**Safety and health program/Organization**			
		A. Policy statement			
		B. Core values			
		C. Facility safety committee/ task group responsibilities			
		1. Safety Activities Task Group			
		2. Rules and Procedures Task Group			
		3. Inspections and Audits Task Group			
		4. Fire and Emergency Task Group			
		5. Education and Training Task Group			
		6. Health and Environmental Task Group			
		7. Accident Investigation Task Group			
		8. Housekeeping Task Group			
		D. Plant Safety and Health Committee(s)			
	II.	**Safety and health rules**			
		A. General plant safety and health rules			
		B. Specific division/department rules			
	III.	**Duties and responsibilities**			
		A. Board of directors			
		B. Chief executive officer, ranking official			
		C. Managers and supervisors			
		D. Employees			
		E. Plant safety–health coordinator			
		F. Employee and supervisory performance appraisals			

Figure A-6. *(continued)*

Check When Complete		Subject	To Be Developed by	Target Date	Date Completed
		SAFETY MANUAL CHECKLIST	Date:		
			Page 3 of 7		
	IV.	**Safety and health training**			
	A.	Employee selection and placement			
	B.	Safety orientation and placement			
	C.	Supervisory safety training			
	D.	Employee safety meetings and contacts			
	E.	Special safety and health training 1. Transfers 2. New assignments			
	V.	**Accident/incident response**			
	A.	Accident reporting			
	B.	Accident investigation			
	C.	Injury and illness recordkeeping			
	D.	Transportation of injured or ill employees			
	E.	Notification of injured employee's family			
	F.	Reporting to government agencies			
	VI.	**Inspection and audit programs**			
	A.	Safety audit and inspection procedure			
	B.	Safety inspection checklists			
	C.	Safety audits of new or modified equipment and facilities			
	D.	Capital project review procedure			
	E.	Monitoring for air contaminants			
	F.	Monitoring for physical health hazards			
	G.	Management of change			
	H.	Preuse analysis procedures			
	VII.	**Effective management action**			
	A.	Tracking of corrective actions			
	B.	Follow-up assessments			
	C.	Safety work orders			

Figure A-6. (*continued*)

Check When Complete		Subject	To Be Developed by	Target Date	Date Completed
		SAFETY MANUAL CHECKLIST	Date:		
			Page 4 of 7		
	VIII.	**Safety permits**			
	A.	Permit systems			
	B.	Confined-space entry			
	C.	Hot-work permit			
	IX.	**Electrical safety**			
	A.	Hazardous locations			
	B.	Temporary electrical wiring			
	C.	Electrical safety work practices for employees			
	X.	**OSHA inspections**			
	XI.	**Contractor and visitor safety and health**			
	XII.	**Safety and health procedures**			
	A.	Acids and caustics			
	B.	Asbestos maintenance and removal			
	C.	Backsiphonage and cross-connection			
	D.	Barricade procedures			
	E.	Biohazards			
	F.	Bloodborne pathogens			
	G.	Chemical hygiene plan			
	H.	Color coding			
	I.	Compressed air			
	J.	Compressed-gas storage and handling			
	K.	Confined-space entry			
	L.	Cranes and hoists			
	M.	Demolition			
	N.	Elevated work			

Figure A-6. *(continued)*

		SAFETY MANUAL CHECKLIST		Date:		
				Page 5 of 7		
Check When Complete		**Subject**	**To Be Developed by**	**Target Date**	**Date Completed**	
	O.	Emergency action plan				
		1. Fire and explosions				
		2. Chemical leaks and spills				
		3. Meteorological occurrences				
		4. Bomb threats/public disorders				
	P.	Employee exposure and medical records policy				
	Q.	Ergonomics				
	R.	Excavating				
	S.	Eyewash stations and safety showers				
	T.	Fall protection				
	U.	Fire protection				
	V.	First aid				
	W.	Flammable and combustible liquids				
	X.	Fleet safety and vehicle maintenance				
	Y.	Forklifts, industrial trucks				
	Z.	Hazard communication program				
	AA.	Hearing conservation program				
	BB.	Helicopter safety				
	CC.	Hot-work procedures				
	DD.	Housekeeping				
	EE.	Industrial hygiene sampling and monitoring				
	FF.	Job safety and health analysis				
	GG.	Ladders, scaffolds, platforms, manlifts				
	HH.	Lead-based paint				

Figure A-6. *(continued)*

Check When Complete		Subject	To Be Developed by	Target Date	Date Completed
		SAFETY MANUAL CHECKLIST	Date:		
			Page 6 of 7		
	II.	Line breaking			
	JJ.	Locking–tagging procedure			
	KK.	Machinery and equipment			
	LL.	Manual material handling			
	MM.	Marine safety			
	NN.	Medical surveillance program			
	OO.	Office safety			
	PP.	Open flames and sparks			
	QQ.	Personal protective equipment			
	RR.	Pipe labeling			
	SS.	Power generation, transmission, and distribution			
	TT.	Pressure vessels			
	UU.	Process hazards			
	VV.	Radiation			
	WW.	Railroad safety			
	XX.	Rescue procedures			
	YY.	Respiratory protection program			
	ZZ.	Rigging			
	AAA.	Tools			
	BBB.	Traffic rules			
	CCC.	Ventilation systems			
	DDD.	Videodisplay terminals			
	EEE.	Walking and working surfaces			
	FFF.	Welding, burning, and cutting			
	GGG.	Work over or adjacent to water			
	HHH.	Work-alone policy			

Figure A-6. *(continued)*

Check When Complete		Subject	To Be Developed by	Target Date	Date Completed
		SAFETY MANUAL CHECKLIST	Date: Page 7 of 7		
	XIII.	**Environmental management policies and procedures**			
	A.	Asbestos disposal			
	B.	Hazardous materials			
	C.	Hazardous waste management			
		1. Generation 2. Treatment 3. Storage 4. Disposal 5. Minimization 6. Pollution prevention			
	D.	Lead disposal			
	E.	Oil spill cleanup and control			
	F.	PCB handling and disposal			
	G.	Pesticide use, handling, and disposal			
	H.	Solid-waste management			
		1. Generation 2. Handling 3. Storage 4. Disposal 5. Minimization 6. Pollution prevention			
	I.	Underground storage tanks			
	J.	Water quality management			
		1. Water treatment 2. Potable water 3. Backsiphonage 4. Cross-connections			

Figure A-6. *(continued)*

SAFETY AND HEALTH MANUAL ASSIGNMENT/AUDIT	Date:
	Page 1 of 3

Company/organization:		Facility:	
Manual assigned to:	**Unit:**		**Custodian title:**

Specific location:

Responsibilities

To ensure that the safety and health manual is current and available for prompt use by personnel, the following responsibilities are assigned to the unit custodian:

- Location: Keep the manual in its assigned location.

- Convenient: Make sure that the manual is convenient for use by all personnel.

- Used: Encourage use of the manual.

- Current: Promptly insert revised rules and procedures and return removed sheets with the verification form.

- Clean: Make sure the manual is clean, neat, and orderly. Obtain new copies as needed.

- Audits: Assist in annual audits of the manual to ensure that it is properly maintained.

Annual audits

Date	Audited by	Assisted by	Date	Audited by	Assisted by

Figure A-7. Sample safety–health manual audit form. *(continued)*

SAFETY AND HEALTH MANUAL ASSIGNMENT/AUDIT		Date:
		Page 2 of 3

Company/organization:		Facility:	
Auditor(s):	Audit date:	Audit time:	Manual number:
Department/unit/group audited:	Specific manual location:	Custodian name and title:	

Audit

Item	Findings (Check One)		Comments (Explain "No" Answers)
	Yes	No	
1. Location: Was the manual in its assigned location?			
2. Convenient: Was the manual located in a convenient place where it could be readily used?			
3. Used: Was there evidence from observation and/or discussion that the manual was being routinely used?			
4. Current: Was the manual current, including all revisions during the past 12 months?			
5. Clean: Was the manual clean, neat, and orderly?			
6. Audit: Was the audit sheet in the manual initialed and dated for this audit?			

Discussions with manual custodian and supervision

1. Suggestions for improving the safety and health manual, including any rules and procedures?

2. Any problems encountered with existing safety–health rules and procedures?

3. Any needed safety–health rules and procedures?

4. Do supervisors know, understand, and follow safety rules and procedures?

Figure A-7. *(continued)*

SAFETY AND HEALTH ASSIGNMENT/AUDIT	Date:
	Page 3 of 3

Additional comments

Recommendations

Audit reviewed by:	Date:

Figure A-7. *(continued)*

CTJ Safety Associates	Training Required by OSHA for General Industry (29 CFR 1910)	Date: May 2003
		Page 1 of 23

Company:	Facility:	Auditor:

Date of audit:	System owner:

GENERAL INFORMATION

This form is a list of the 29 CFR 1904 and 1910 OSHA general industry standards that require training and education. Employees should receive appropriate training at the time of initial job assignment and periodically as needed or required. Training may be necessary for employees who are reassigned; whenever new equipment, procedures, or chemicals are introduced; or to improve current employee work habits. It is important that a record of all training be maintained.

In addition, all ANSI, NFPA, ASME, Compressed Gas Association, etc. standards that are incorporated by reference by to OSHA standards and State OSHA standards must be consulted for training requirements. This list does not include those requirements.

Note: This checklist is current as of May 2003. Updates are available from CTJ Safety.

INDEX

1910.1020	Access to Records	1910.218	Forging Machines
1910.38	Action Plans	1910.1048	Formaldehyde
1910.1045	Acrylonitrile	1910.1200	Hazard Communication
1910.67	Aerial Lifts	1910.120	Hazardous Wastes
1910.165	Alarm Systems	1910.95	Hearing Conservation
1910.111	Ammonia	1910.183	Helicopters
1910.1018	Arsenic	1910.103	Hydrogen
1910.1001	Asbestos	1910.1018	Inorganic Arsenic
1910.263	Bakery	1910.96	Ionizing Radiation
1910.1028	Benzene	1910.1450	Laboratories
1910.1030	Bloodborne Pathogens	1910.264	Laundry
1910.66	Building Maintenance Platforms	1910.1025	Lead
1910.1051	1,3-Butadiene	1910.103	Liquefied Hydrogen
1910.1027	Cadmium	1910.147	Lockout/Tagout
1910.1003-.1016	Carcinogens	1910.266	Logging
1910.1029	Coke Ovens	1910.110	LP Gas

Figure A-8. OSHA requirements for Education and Training Task Group. *(continued)*

INDEX			
1910.101	Compressed Gas Cylinders	1910.68	Manlifts
1910.146	Confined Spaces	1910.217	Mechanical Power Presses
1910.1043	Cotton Dust	1910.1052	Methylene Chloride
1910.179-.180	Cranes	1910.1050	Methylenedianiline
1910.1044	DBCP	1910.95	Noise
1910.181	Derricks	1910.132	Personal Protective Equipment
1910.122-.126	Dip Tanks	1910.217	Power Presses
1910.410	Diving	1910.178	Powered Industrial Trucks
1910.332, 303-.305	Electrical	1910.66	Powered Platforms
1910.269	Electric Power Generation, Transmission, and Distribution	1910.119	Process Safety Management
1910.38	Emergency Action Plans	1910.261	Pulp, Paper, Paperboard
1910.120	Emergency Response	1904.35	Reporting Injuries
1910.1047	Ethylene Oxide	1910.134	Respirators
1910.109	Explosives	1910.265	Sawmills
1910.156	Fire Brigades	1910.28	Scaffolds
1910.157	Fire Extinguishers	1910.145	Signs & Tags
1910.164	Fire Detection Systems	1910.184	Slings
1910.38	Fire Prevention Plans	1910.158	Standpipe & Hose
1910.151	First Aid	1910.268	Telecommunications
1910.161	Fixed Dry Chemical	1910.142	Temporary Labor Camps
1910.160	Fixed Extinguishing	1910.1017	Vinyl Chloride
1910.162	Fixed Gaseous	1910.252-.255	Welding
1910.106	Flammable Liquids	1910.177	Wheel Rims
1910.178	Forklifts	1910.213	Woodworking Machines

Figure A-8. *(continued)*

CTJ Safety Associates	Training Required by OSHA for General Industry (29 CFR 1910)		Date: May 2003
			Page 3 of 23
Standard	**Subject**	**Coverage**[a]	**Frequency**[b]	**Comments**
1904.35 (a)(1)	OSHA recordkeeping (reporting injuries and illnesses)	All employees	Upon assignment	
1910.28 (c)(6)	Tube and coupler scaffolds (erected by competent and experienced personnel)	Involved personnel	Upon assignment	
1910.28 (d)(12)	Tubular welded frame scaffolds (erected by competent and experienced personnel)	Involved personnel	Upon assignment	
1910.28 (f)(17)	Masons adjustable multipoint suspension scaffold (installed or relocated by competent person)	Designated personnel	Upon assignment	
1910.28 (h)(10)	Stone setters adjustable multipoint suspension scaffold (installed or relocated by competent person)	Designated personnel	Upon assignment	
1910.38 (a)(5)	Emergency action plans (employee evacuation)	All employees	Upon initial or change of assignment/ plan change	
1910.38 (b)(4)	Fire prevention plans (fire hazards of materials and processes)	All employees	Upon initial or change of assignment/ plan change	
1910.66 (e)(9)	Building maintenance powered platforms (escape routes, emergency procedures, alarm systems)	Exposed employees	Upon assignment/ plan change	
1910.66(g)	Building maintenance powered platforms (inspector competency)	Inspectors	Upon assignment	
1910.66 (i)(1)	Building maintenance powered platforms (hazard recognition, emergencies, work procedures, personal fall arrest systems)	Operators	Upon assignment	
1910.66 (i)(1)(iii)	Building maintenance powered platforms (training by competent person)	Trainers	Upon assignment	

Figure A-8. *(continued)*

CTJ Safety Associates	Training Required by OSHA for General Industry (29 CFR 1910)		Date: May 2003
			Page 4 of 23
Standard	**Subject**	**Coverage**	**Frequency**	**Comments**
1910.67 (c)(2)(ii)	Extensible and articulating aerial lifts (operator competency)	Operators	Upon assignment	
1910.68 (b)(1)	Manlifts (trained in their use)	Users	Upon assignment	
1910.68 (e)(1)	Manlifts (inspections by a competent designated person)	Designated individual	Upon assignment	
1910.95 (k)(1)	Occupational noise exposure (hearing conservation, hazards of exposure, protective devices, regulations)	Employees exposed to TWA > 85 dBA	Upon initial assignment/ annual retraining	
1910.96 (f)(3)(viii)	Ionizing radiation evacuation warning signals (testing and familiarity with signals, drills)	Employees normally in area	Upon assignment	
1910.96 (i)(2)	Ionizing radiation instruction of personnel and posting (anyone in area to be instructed in problems, precautions, regulations, reports of exposure)	Anyone in area	Upon entering area	
1910.101 (b)	Compressed-gas cylinders (those who handle cylinders)	Employees who handle, store, or use cylinders	Upon assignment	
1910.103 (c)(4)(ii)	Liquefied hydrogen systems (unloading hydrogen-supply-qualified person in attendance)	Qualified person	Upon assignment	
1910.106 (b)(5)(vi)	Flammable liquids (tanks stored in flood areas; operators to be thoroughly informed and detailed instructions)	Operators	Upon assignment	
1910.109 (d)(1)(iii)	Transportation of explosives (informing local fire and police, handling under qualified supervision)	Involved personnel	Upon assignment	

Figure A-8. *(continued)*

CTJ Safety Associates	Training Required by OSHA for General Industry (29 CFR 1910)		Date: May 2003
			Page 5 of 23
Standard	**Subject**	**Coverage**	**Frequency**	**Comments**
1910.109 (d)(2)(iii) (*b*)	Transportation vehicles-explosives (inspection of fire extinguishers)	Inspectors	Upon assignment	
1910.109 (d)(3)(i)	Operation of vehicles-explosives (familiar with regulations and laws)	Drivers	Upon assignment	
1910.109 (d)(3)(iii)	Operation of vehicles-explosives (attend vehicle at all times, aware of materials, dangers, measure and procedures to protect the public)	Driver or attendant	Upon assignment	
1910.109 (g)(3)(iii) (*a*)	Transportation and handling of blasting agents (operator training, system evaluator qualifications, driver, DOT standards)	Operators, evaluator, driver	Upon assignment	
1910.109 (g)(6)(ii)	Transportation of blasting agents (drivers of motor vehicles familiar with traffic laws)	Drivers	Upon assignment	
1910.109 (h)(4)(ii)(*b*)	Delivery of water gels (drivers of motor vehicles trained in vehicle operations, emergencies)	Drivers	Upon assignment	
1910.109 (i)(6)(iii)	Ammonium nitrate storage areas (authorized entrants trained)	Entrants	Upon assignment	
1910.110 (b)(16)	Storage and handling of liquified petroleum gases (installation, removal, operation, maintenance)	Anyone performing operation	Upon assignment	
1910.110 (d)(12)(i)	Liquified petroleum gases (non-DOT containers watch service by trained personnel)	Associated personnel	Upon assignment	
1910.110 (h)(11)(vii)	LPG service stations (dispensing only by a competent attendant)	Attendant	Upon assignment	

Figure A-8. *(continued)*

CTJ Safety Associates	Training Required by OSHA for General Industry (29 CFR 1910)		Date: May 2003
			Page 6 of 23
Standard	**Subject**	**Coverage**	**Frequency**	**Comments**
1910.111 (b)(1)(iv)	Storage and handling of anhydrous ammonia (equipment and systems safety evaluation for custom design and construction by PE or other)	Evaluator	Before evaluation	
1910.111 (b)(13)(ii)	Ammonia (unloading by reliable, properly instructed persons to monitor compliance with applicable procedures)	Monitoring and operating personnel	Upon assignment	
1910.119 (g)(1)	Operations of processes of highly hazardous chemicals	Operators	Initially every 3 years	
1910.119 (h)(2)(ii)	Contractors at sites with processes of highly hazardous chemicals	Contractors	Upon assignment	
1910.119 (h)(3)	Contract employees at sites with processes of highly hazardous chemicals	Each contract employee	Upon assignment	
1910.119 (j)(3)	Process maintenance of highly hazardous chemicals	Maintenance workers	Upon assignment	
1910.119 (l)(3)	Management of change of processes of highly hazardous chemicals	Workers affected by a change of process	Prior to startup	
1910.120 (b)(1)(iv)	Hazardous-waste operations	Contractors and subcontractors	Prior to any site activity	
1910.120 (b)(4)(iii)	Hazardous-waste operations (preentry briefing)	Exposed employees	Prior to site entry	
1910.120 (c)(8)	Hazardous-waste operations (properties of known substances)	Exposed employees	Prior to start of activities	
1910.120 (e)(3)	Hazardous-waste operations	Exposed employees	Initially annually	
1910.120(i)	Hazardous-waste operations (nature, level, and degree of exposure)	Exposed employees, contractors, and subcontractors	Prior to exposure	

Figure A-8. *(continued)*

CTJ Safety Associates	Training Required by OSHA for General Industry (29 CFR 1910)		Date: May 2003	
			Page 7 of 23	
Standard	**Subject**	**Coverage**	**Frequency**	**Comments**
1910.120 (j)(1)(vi)	Hazardous-waste operations (movement of drums or containers—potential hazards)	Exposed employees	Prior to movement	
1910.120 (k)(7)	Hazardous-waste operations (harmful effects of contaminated clothing)	Commercial laundries and cleaners	Prior to exposure	
1910.120 (p)(7)	TSD operations	Exposed employees	Upon assignment/ annually	
1910.120 (p)(8)(iii) (A)–(C)	Emergency response training	Emergency response employees at RCRA TSD facilities	Before assignment	
1910.120 (q)(6),(8)	Emergency response to hazardous-substance releases	Exposed employees	Upon assignment/ annually	
1910.120 (q)(7)	Emergency response to hazardous-substance releases	Trainers	Upon assignment	
1910.124(f)	Dip tanks (first aid)	Exposed employees	Upon assignment	
1910.132(f)	PPE (eye, face, head, foot, hand)	Affected employees	Initial, changes, inadequacies	
1910.134 (c)(3)	Respiratory protection program	Program administrator	Before assignment	
1910.134 (h)(4)(i)	Respirators (repair on respirators)	Respirator repair techician	Before assignment	
1910.134 (k)(1)(i)	Respirators (necessity of respirator; proper use, fit and maintenance)	Respiratory user	Upon assignment; annually; changes; employee need	

Figure A-8. *(continued)*

CTJ Safety Associates	Training Required by OSHA for General Industry (29 CFR 1910)		Date: May 2003
			Page 8 of 23
Standard	**Subject**	**Coverage**	**Frequency**	**Comments**
1910.134 (k)(1)(ii)	Respirators (limitations and capabilities)	Respirator user	Upon assignment; annually; changes; employee need	
1910.134 (k)(1)(iii)	Respirators (use in emergencies)	Respirator user	Upon assignment; annually; changes; employee need	
1910.134 (k)(1)(iv)	Respirators (how to inspect, put on, use, and check seals)	Respirator user	Upon assignment; annually; changes; employee need	
1910.134 (k)(1)(v)	Respirators (maintenance and storage)	Respirator user	Upon assignment; annually; changes; employee need	
1910.134 (k)(1)(vi)	Respirators (recognition of medical signs and symptoms that may limit or prevent effective use)	Respirator user	Upon assignment; annually; changes; employee need	
1910.134 (k)(1)(vii)	Respirators (general requirements)	Respirator user	Upon assignment; annually; changes; employee need	
1910.134 (k)(2)	Respirators (understandable training)	Respirator user	Upon assignment; annually; changes; employee need	

Figure A-8. *(continued)*

CTJ Safety Associates	Training Required by OSHA for General Industry (29 CFR 1910)		Date: May 2003
			Page 9 of 23
Sttandard	**Subject**	**Coverage**	**Frequency**	**Comments**
1910.134 (k)(3)	Respirators (initial training)	Respirator user	Before using the respirator	
1910.134 (k)(4)	Respirators (prior training for new employees)	Respirator user	Before beginning assignment; annually from previous training date; changes; employee need	
1910.134 (k)(5)	Respirators (retraining)	Respirator user	Annually; changes; inadequacies; employee need	
1910.142 (k)(2)	Temporary labor camp	First aid personnel	Upon assignment	
1910.145 (c)(1)(ii)	Danger signs and tags (immediate danger and precautions to be taken)	All employees	Upon assignment	
1910.145 (c)(2)(ii)	Caution signs and tags (possible hazards and precautions to be taken)	All employees	Upon assignment	
1910.145 (f)(4)(v)	Accident prevention signs and tags (employees instructed in the meaning and special precautions)	All employees	Upon assignment	
1910.146 (c)(2)	Confined spaces (inform employees about permit spaces)	Exposed employees	Upon assignment	
1910.146 (c)(8)(i)–(v)	Confined spaces (host employer must apprise contractor about permit spaces)	Contractors	Upon arrangement of work	

Figure A-8. *(continued)*

CTJ Safety Associates	Training Required by OSHA for General Industry (29 CFR 1910)		Date: May 2003
			Page 10 of 23
Standard	**Subject**	**Coverage**	**Frequency**	**Comments**
1910.146 (g),(h),(i) and (j)	Confined spaces (knowledge and skills to safely work in permit spaces)	Entrants, attendants, entry supervisors, contractors	Before assignment; change in assignment; change in space; a new hazard; inadequacies in employees' knowledge and use of procedures	
1910.146 (k)(2)(i)	Confined spaces (rescue and emergency services)	Employees designated to provide rescue and emergency services trained in the use of the PPE needed to perform duties	Initially	
1910.146 (k)(2)(ii)	Confined spaces (rescue and emergency services)	Employees designated to provide rescue and emergency services trained on assigned duties	Initially	
1910.146 (k)(2)(iii)	Confined spaces (rescue and emergency services)	Employees designated to provide rescue and emergency services trained in first aid and CPR	Initially	
1910.146 (k)(2)(iv)	Confined spaces (rescue and emergency services)	Employees designated to provide rescue and emergency services trained with practice rescues	At least annually	
1910.147 (c)(6)(i)(D)	Lockout/tagout (employee's responsibilities when using tags)	Employees who use tags	Initially/ Annually	
1910.147 (c)(7)	Lockout/tagout (training for control of hazardous energy)	Authorized and affected employees	Prior to start of work	
1910.147 (f)(2)	Lockout/tagout (inform contractors of program and vice versa)	Contractors	Upon assignment	

Figure A-8. *(continued)*

CTJ Safety Associates	Training Required by OSHA for General Industry (29 CFR 1910)		Date: May 2003
			Page 11 of 23
Standard	**Subject**	**Coverage**	**Frequency**	**Comments**
1910.151 (b)	Medical services and first-aid (availability of medical personnel and in absence, trained first-aiders, supplies approved by physician)	First aid personnel	Upon assignment	
1910.156 (c)	Fire brigades (training on emergency activities, special hazards, special substances and precautions, change in status, functions of members, etc.)	Fire brigade members, instructors, and leaders (special)	Before assignment/ annual retraining, certain quarterly training	
1910.157 (d)(3)	Standpipes (use of uniformly spaced standpipes in lieu of extinguishers requires employee training)	Employee designated to use	Upon assignment/ annual retraining	
1910.157(f)	Portable fire extinguishers (hydrostatic testing to be performed by trained personnel)	Testing personnel	Upon assignment	
1910.157 (g)	Portable fire extinguishers (training on use, fire principles, hazards, emergency action plans)	Employees who may use equipment	Upon initial assignment/ annual retraining	
1910.158 (e)(2)(vi)	Standpipe–hose system (testing, maintenance, and inspections to be performed by trained personnel)	Inspectors, testing personnel	Upon assignment	
1910.160 (b)(2), (6)	Fixed extinguishing systems (notification if system is inoperable, warnings of hazardous-agent atmospheres, warning or caution signs, system inspected by knowledgeable person)	Employees working in area	Upon assignment	
1910.160 (b)(10)	Fixed systems (inspect, maintain, operate)	Designated employees	Upon assignment/ annual	

Figure A-8. *(continued)*

| CTJ Safety Associates | Training Required by OSHA for General Industry (29 CFR 1910) | | Date: May 2003 |
| | | | Page 12 of 23 |
Standard	Subject	Coverage	Frequency	Comments
1910.161 (b)(3)	Fixed dry chemical extinguishing system (predischarge alarm to be used)	Employees in area	Upon assignment	
1910.162 (b)(5)	Fixed gaseous extinguishing agent (predischarge alarm to be used)	Employees in area	Upon assignment	
1910.164 (c)(4)	Fire detection systems (maintenance, testing, and adjustments to be performed by trained personnel)	Testing and maintenance personnel	Upon assignment	
1910.165 (b)(4)	Employee alarm systems (reporting emergencies, posting numbers, etc., emergency alarms)	Each employee	Upon assignment	
1910.165 (d)(5)	Employee alarm systems (maintenance, testing, servicing performed by trained persons)	Testing, maintenance, and servicing personnel	Upon assignment	
1910.177 (c)	Single-piece and multipiece rim wheels	Servicing employees	Upon assignment/ as necessary	
1910.178 (l)(1) and (4)	Powered industrial trucks (operator training required)	All operators	Initially/ refresher training when • Operating vehicle in unsafe manner • Operator is involved in an accident or near-miss accident • An unsafe operation evaluation • Assignment to a different truck • Changes in the workplace	

Figure A-8. *(continued)*

CTJ Safety Associates	Training Required by OSHA for General Industry (29 CFR 1910)		Date: May 2003
			Page 13 of 23
Standard	**Subject**	**Coverage**	**Frequency**	**Comments**
1910.178 (l)(7)	Powered industrial truck (operator training required)	All operators hired by December 1, 1999	Training conducted before December 1, 1999	
1910.179 (b)(8)	Overhead and gantry cranes (designated, qualified persons required to operate)	Operators	Before assignment	
1910.179 (l)(3)(i) & (iii)(a)	Cranes (maintenance, adjustments performed by qualified person, competent supervisor)	Maintenance and supervisory personnel	Upon assignment	
1910.179 (n)(3)(ix)	Cranes (use of two or more cranes to lift requires one qualified person to supervise, instruct all personnel inoperation)	Supervisor	Upon assignment	
1910.179 (o)(3)	Cranes (operators to be familiar with fire extinguishers)	Operators	Upon assignment/ annual	
1910.180 (b)(3)	Crawler locomotive and truck cranes (only qualified persons may operate)	Operators	Before assignment	
1910.180 (h)(3)(xii)	Cranes (use of two or more cranes to lift; requires one qualified person to supervise, instruct all personnel in operation)	Supervisor	Upon assignment	
1910.180 (i)(5)(ii)	Cranes (operators to be familiar with use of fire extinguishers)	Operators	Upon assignment/ annual	
1910.181 (b)(3),(h), (j)(3)(ii)	Derricks (training required for operation and supervision, and fire extinguishers)	Operators, supervisors	Upon assignment	
1910.183 (b),(d),(m), (n),(p)	Helicopters (briefings, cargo hooks tested, signal systems, visibility, approaching helicopters)	Designated personnel	Upon assignment	

Figure A-8. *(continued)*

| CTJ Safety Associates | Training Required by OSHA for General Industry (29 CFR 1910) | | Date: May 2003 |
| | | | Page 14 of 23 |

Standard	Subject	Coverage	Frequency	Comments
1910.184 (d)	Slings (daily inspections by competent person)	Designated person	Upon assignment	
1910.184 (e)(3)(iii)	Slings (inspection of alloy steel chain slings by a competent person)	Designated person	Upon assignment	
1910.213 (s)(5)	Woodworking machinery (sharpening and tensioning of saw blades or cutters by competent persons)	Operators	Upon assignment	
1910.217 (e)(3)	Mechanical power presses (maintenance, modifications, and inspections by competent persons)	Maintenance personnel	Upon assignment/ ensure continued competence	
1910.217 (f)(2) & (h)(13)	Power presses (operators to be trained in safe work methods; adequate supervision)	Operators	Upon assignment	
1910.217 (h)(5)(ii)	Power presses (brake monitors adjusted by authorized person)	Designated person	Upon assignment	
1910.217 (h)(9)(vii)	Power presses (PSDI adjusted by authorized person)	Designated person	Upon assignment/ periodic	
1910.217 (h)(10)(vi)	Power presses (competent personnel to inspect and maintain power presses with PSDI)	Designated person	Upon assignment/ periodic	
1910.217 (h)(12)(iii)	Power presses (die change checked by authorized person)	Designated person	Upon assignment	
1910.218 (a)(2)(iii)	Forging machines (training for proper inspection and maintenance)	Assigned personnel	Before assignment	
1910.252 (a)(2) (iii)(B)	Welding: general (fire watcher to be trained in use of fire extinguishing equipment; familiar with alarms)	Fire watcher	Upon assignment	

Figure A-8. *(continued)*

CTJ Safety Associates	Training Required by OSHA for General Industry (29 CFR 1910)		Date: May 2003
			Page 15 of 23
Standard	**Subject**	**Coverage**	**Frequency**	**Comments**
1910.252 (a)(2) (xiii)(C)	Welding: general (management to insist that cutters or welders be properly trained in safe practice)	Management, welders, cutters	Upon assignment	
1910.252 (c)(13)	Welding: general (first-aid-trained employees to be available on every shift, when welding)	First-aiders	Upon assignment	
1910.253 (a)(4)	Welding and cutting: oxygen fuel gas (workers in charge of fuel gas supply equipment)	Workers in charge	Before being left in charge	
1910.253 (c)(5)(i)	Welding and cutting: oxygen–fuel gas	Installation of cylinder manifolds	Before assignment	
1910.253 (e)(6)(ii)	Welding and cutting: oxygen–fuel gas (properly instructed mechanics to repair regulators)	Assigned personnel	Before assignment	
1910.254 (a)(3)	Welding: arc welding and cutting (employee to be instructed and qualified in operation of equipment as specified in regulations)	Operators (welders)	Before assignment	
1910.254 (d)(1)	Welding: arc welding and cutting (all welders are required to be acquainted with requirements of regulations; operators to report defects; repairs to be made by qualified personnel)	Welders, maintenance, personnel	Upon assignment	
1910.255 (a)(1),(3)	Welding: resistance (resistance welders installed by qualified electrician to conform with regulations; welders to be properly instructed and judged competent to operate)	Electricians, welders, supervisors	Before assignment	

Figure A-8. *(continued)*

| CTJ Safety Associates | Training Required by OSHA for General Industry (29 CFR 1910) | | Date: May 2003 |
| | | | Page 16 of 23 |

Standard	Subject	Coverage	Frequency	Comments
1910.255 (e)	Welding: resistance (inspection and maintenance by qualified person)	Maintenance personnel	Upon assignment	
1910.261 (c)(14)(ii)	Pulp, paper, and paperboard (dynamite handled by authorized personnel)	Designated individual	Upon assignment	
1910.261 (g)(15)(iii)	Pulp, paper, and paperboard (experienced employee stationed outside digester during entry)	Designated individual	Upon assignment	
1910.263 (l)(9)(ii)	Bakery equipment (employees instructed to inspect ovens)	Designated individual	Upon assignment	
1910.264 (d)(1)(v)	Laundry machinery (employees instructed as to hazards and safe practices)	All employees	Upon assignment	
1910.265 (c)(24)(iii)	Sawmills (qualified person to inspect hoist, ropes, cables, slings, chains)	Designated individual	Upon assignment	
1910.266 (d)(7)(iii)	Logging (signaling)	Designated personnel	Upon assignment	
1910.266 (d)(10)(ii)	Logging (explosives and blasting agents)	Designated personnel	Upon assignment	
1910.266 (f)(2)(i)	Logging (machine operation)	Designated personnel	Upon assignment	
1910.266 (i)(1)-(6)	Logging (work tasks, tools, machines, procedures, practice, standard)	Each employee and supervisor	Upon assignment	
1910.266 (i)(7)	Logging (first aid and CPR)	Each employee and supervisor	Upon assignment; first aid every 3 years; CPR annually	

Figure A-8. *(continued)*

CTJ Safety Associates	Training Required by OSHA for General Industry (29 CFR 1910)		Date: May 2003	
			Page 17 of 23	
Standard	**Subject**	**Coverage**	**Frequency**	**Comments**
1910.266 (i)(8)	Logging (competent trainer)	Trainer	Upon assignment	
1910.266 (i)(9)	Logging (training understandable for employees)	All employees	Each training session	
1910.266 (i)(11)	Logging (safety and health meetings)	Each employee	Monthly	
1910.268 (c)	Telecommunications (safe practices)	All employees	Upon assignment	
1910.268 (e)	Telecommunications (tools and protective equipment inspected by competent person)	Designated individual	Upon assignment	
1910.268 (g)(3)(ii)	Telecommunications (pole climbers inspected by competent person)	Designated individual	Upon assignment	
1910.268 (h)(1)	Telecommunications (ladders inspected by competent person)	Designated individual	Upon assignment	
1910.268 (j)(1)	Telecommunications (vehicle material handling devices inspected by competent person)	Designated individual	Upon assignment	
1910.268 (j)(4)(iv)(D)	Telecommunications (derrick operators trained)	Derrick operators	Upon assignment	
1910.268 (j)(4)(iv)(F)	Telecommunications (derricks inspected by competent person)	Designated individual	Upon assignment	
1910.268 (l)(1)	Telecommunications (employees trained to locate and test cable faults)	Designated individual	Upon assignment	
1910.268 (q)	Telecommunications (employees trained in tree trimming)	Designated individual	Upon assignment	
1910.269 (a)(2)	Electric power generation, transmission, and distribution (work practices, procedures and requirements of standard)	All employees	Upon assignment; new procedures, technology	

Figure A-8. *(continued)*

CTJ Safety Associates	Training Required by OSHA for General Industry (29 CFR 1910)		Date: May 2003
			Page 18 of 23

Standard	Subject	Coverage	Frequency	Comments
1910.269 (b)(1)	Electric power generation, transmission, and distribution (first aid/CPR-trained when exposed to ≥50 V)	Employees exposed to lines or equipment.	Upon assignment	
1910.269 (c)	Electric power generation, transmission, and distribution (job briefing before each job)	Work crew	Before each job	
1910.269 (d)(2)(vi)	Electric power generation, transmission, and distribution (energy control procedures)	All authorized and affected employees, and all other employees	Upon assignment; changes	
1910.269 (d)(2)(vii)	Electric power generation, transmission, and distribution (use of tagout systems for energy control)	All authorized and affected employees, and all other employees	Upon assignment; changes	
1910.269 (d)(8)(iv)	Electric power generation, transmission, and distribution (contractor energy control procedures)	All contractors and host employees	Upon assignment	
1910.269 (e)(2)	Electric power generation, transmission, and distribution (enclosed space entry and rescue)	Those who enter and attendants	Upon assignment	
1910.269(f)	Electric power generation, transmission, and distribution (excavation)	Competent persons	Upon assignment	
1910.269 (l)(1)	Electric power generation, transmission, and distribution (working on or near exposed live parts)	Qualified employees	Upon assignment	
1910.269 (l)(6)(ii)	Electric power generation, transmission, and distribution (hazards of flames or electric arcs)	Exposed employees	Upon assignment	
1910.269 (o)(2)(ii)	Electric power generation, transmission, and distribution (safe work practices for testing and test facilities)	Exposed employees	Upon assignment	

Figure A-8. *(continued)*

CTJ Safety Associates	Training Required by OSHA for General Industry (29 CFR 1910)		Date: May 2003
			Page 19 of 23
Standard	**Subject**	**Coverage**	**Frequency**	**Comments**
1910.269 (q)(3)(i)	Electric power generation, transmission, and distribution (live-line bare-hand work)	Exposed employees	Upon assignment; new procedure, technology	
1910.269 (r)(1)(vi)	Electric power generation, transmission, and distribution (line clearance tree trimming operations in the aftermath of an emergency)	Exposed employees	Upon assignment	
1910.269 (t)(3)(i)	Electric power generation, transmission, and distribution (first aid/CPR at manholes)	Attendant	Upon assignment	
1910.269 (v)(6)(i)	Electric power generation, transmission, and distribution (water or steam spaces in power generation)	Designated inspector of conditions	Upon assignment	
1910.269 (v)(8)(ii)	Electric power generation, transmission, and distribution (chlorine systems in power generation)	Designated employees who enter restricted areas	Upon assignment	
1910.269 (v)(11)(i)	Electric power generation, transmission, and distribution (railroad equipment at coal and ash handling)	Designated persons	Upon assignment	
1910.269 (v)(11)(vii)	Electric power generation, transmission, and distribution (coal or ash handling conveyor belts)	Employees who work in this area	Upon assignment	
1910.272 (e)	Grain handling facilities (safety precautions, procedures, special tasks)	All employees	Upon assignment; annually; job changes	
1910.272 (g)(5)	Grain handling facilities (observer of grain bin entry rescue procedures)	Observer	Upon assignment	

Figure A-8. *(continued)*

| CTJ Safety Associates | Training Required by OSHA for General Industry (29 CFR 1910) | | Date: May 2003 |
			Page 20 of 23
Standard	**Subject**	**Coverage**	**Frequency**	**Comments**
1910.272(i)	Grain handling facilities (contractors informed of hazards and emergency plans)	Contractors	Upon assignment	
1910.303 (g)(2)(i)(A), (h)(2)(i)	Electrical system (installation accessible by qualified persons)	Electricians	Upon assignment	
1910.304 (d)(2)(i)	Electrical services (service entrance conductors accessible to qualified personnel)	Electricians	Upon assignment	
1910.304 (f)(1)(v) (c)(2)	Electrical grounding (qualified person to service AC system 50–1,000 V)	Electricians	Upon assignment	
1910.305 (h)	Portable cables over 600 V (terminations accessible only to qualified personnel)	Electricians	Upon assignment	
1910.305 (j)(4)(iv) (a)(1)	Electrical motors (some motors accessible to qualified persons)	Electricians	Upon assignment	
1910.332	Safety-related work practices (employees at risk of electrical shock)	Designated occupations/ all employees	Upon assignment	
1910.410 (a) & (b)	Diving operations	Dive team members	Upon assignment	
1910.1001 (j)(2)	Asbestos (inform employees and employer about ACM and PACM)	Employers and building owners	Initially	
1910.1001 (j)(3)(iv)	Asbestos (understanding warning signs)	Employees who work in and contiguous to regulated areas	Upon assignment	
1910.1001 (j)(7)(i)-(iii)	Asbestos	Exposed employees to PEL or excursion limit	Upon assignment; annually	
1910.1001 (j)(7)(iv)	Asbestos (awareness) of ACM and PACM	Housekeeping personnel	Upon assignment; annually	

Figure A-8. *(continued)*

CTJ Safety Associates	Training Required by OSHA for General Industry (29 CFR 1910)		Date: May 2003
			Page 21 of 23
Standard	**Subject**	**Coverage**	**Frequency**	**Comments**

Standard	Subject	Coverage	Frequency	Comments
1910.1003-.1016	Carcinogens—regulated areas (training in toxicity, hazards, exposure operations, medical surveillance, self-exam and symptoms, decontamination, emergency procedures, etc.) 4-Nitrobiphenyl α-Naphthylamine Methyl chloromethyl ether 3,3′-Dichlorobenzidine (and its salts) Bis-(Chloromethyl ether) β-Naphthylamine Benzidine 4-Aminodiphenyl Ethyleneimine β-Propiolactone 2-Acetylaminofluorene 4-Dimethylaminoazobenzene N-Nitrosodimethylamine	Any employee entering area	Upon assignment; annual retraining	
1910.1017 (j)	Vinyl chloride	Exposed employees	Indoctrination; annually	
1910.1018 (o)	Inorganic arsenic	Employees exposed above action level (AL)	Initial assignment; quarterly, annually	
1910.1020 (g)	Access to employee exposure and medical records	Employees exposed to toxic chemicals and harmful physical agents	Upon assignment; annually	
1910.1025 (l)(1)	Lead	All exposed employees	Initial assignment; annually	
1910.1027 (m)	Cadmium	All exposed employees	Upon assignment; annually	
1910.1028 (j)(3)	Benzene	Exposed employees	Initial assignment; annually if exposed above AL	

Figure A-8. *(continued)*

CTJ Safety Associates	Training Required by OSHA for General Industry (29 CFR 1910)		Date: May 2003
			Page 22 of 23
Standard	**Subject**	**Coverage**	**Frequency**	**Comments**

Standard	Subject	Coverage	Frequency	Comments
1910.1029 (k)	Coke oven emissions	Employees in the regulated area	Initial assignment; annually	
1910.1030 (e)(5)	HIV and HBV research and production facilities (microbiological practices)	All covered employees	Initial assignment; annually	
1910.1030 (g)(2)	Bloodborne pathogens (epidemiology, symptoms, modes of transmission, control procedures, decontamination, HBV vaccine)	All covered employees	Initial assignment; annually	
1910.1030 (g)(2)(viii)	Bloodborne pathogens (trainer knowledgeable)	Trainers	Upon assignment	
1910.1043 (i)	Cotton dust	All exposed employees	Prior to initial assignment; annually; work changes; when need for retraining is indicated	
1910.1044 (n)	1,2-Dibromo-3-chloropropane	All exposed employees	Upon assignment	
1910.1045 (o)	Acrylonitrile	All exposed employees	Initial assignment; annually	
1910.1047 (j)(3)	Ethylene oxide	Employees exposed above AL	Initial assignment; annually	
1910.1048 (n)	Formaldehyde	Employee exposed above AL or STEL	Prior to assignment; annually	
1910.1050 (k)	Methylenedianiline	All exposed employees	Initial assignment; annually	
1910.1051 (e)(4)	1,3-Butadiene	All exposed employees	Upon assignment	
1910.1051 (l)(2)	1,3-Butadiene	All exposed employees	Initial assignment	

Figure A-8. *(continued)*

CTJ Safety Associates	Training Required by OSHA for General Industry (29 CFR 1910)		Date: May 2003
			Page 23 of 23
Standard	**Subject**	**Coverage**	**Frequency**	**Comments**

Standard	Subject	Coverage	Frequency	Comments
1910.1052 (e)(7)	Methylene chloride—communicate the access restrictions and locations of regulated areas	All other employers with work operations at that worksite	Initial assignment; repeated as necessary	
1910.1052 (k)	Methylene chloride—hazard communication	All exposed employees	Initial assignment; workplace changes; repeated as necessary	
1910.1052 (l)(1)-(l)(3)	Methylene chloride (MC)	All affected employees provided with information, training, and requirements of section (l) and MC's appendixes	Prior to or at initial assignment; workplace changes; repeated as necessary	
1910.1052 (l)(7)	Methylene chloride—hazard communication, multiemployer worksite	Employees exposed to methylene chloride	Initial assignment; workplace changes; repeated as necessary	
1910.1200 (h)	Hazard communication: provide information about hazardous chemicals; standards, operations, physical and health hazards, protective measures, target organ(s)	Any employee exposed to hazardous chemicals	Upon assignment, and whenever a new hazard is introduced	
1910.1450 (f)	Laboratories: employees apprised of chemical hazards in the work area	All exposed employees	Initial assignment; new exposures; repeated as necessary	
1910.1450 (h)(2)(i)	Laboratories: chemicals produced exclusively for the laboratory	All exposed employees	Initial assignment; repeated as necessary	

[a] In general, implied training through the use of signs, tags, warning devices, and notifications of employees was not covered.
[b] "Upon assignment" was used when initial training was not designated.

Figure A-8. *(continued)*

SAFETY MEETING AUDIT		Date:	
		Page 1 Of 1	

Organization	Location	Department
Meeting date	**Meeting time**	**Meeting location**
Instructor(s)		**Instructor(s) title(s)**
Meeting subject(s)		

Audit

No.	Item	Check		Comments
		Check		**Comments**
No.	**Item**	**Yes**	**No**	**(Explain "No" Answers)**
1	Good meeting location?			
2	Good time for meeting?			
3	Meeting facilities and equipment proper?			
4	Pertinent meeting subject(s)?			
5	Meeting purpose communicated?			
6	Instructor well prepared?			
7	Meeting attendees interested?			
8	Attendees participated?			
9	Attendees asked for opinions and suggestions?			
10	Meeting summary given?			
11	Meeting record kept?			
12				
13				
14				
Auditor's signature:		**Title:**		**Date:**

Figure A-9. Sample safety meeting audit form.

CTJ Safety Associates	Inspections, Audits, Tests, and Monitoring Required by OSHA for General Industry (29 CFR 1910)	Date: May 2003
		Page 1 of 23

Company:	Facility:	Auditor:
Date of audit:	System owner:	

GENERAL INFORMATION

This form is a list of the 29 CFR 1910 OSHA general industry standards that requires inspections, tests, and audits of workplaces and equipment, and monitoring of chemicals and physical health hazards. Employees who conduct these inspections and audits should be thoroughly trained and well motivated. A timetable that complies with the OSHA standards or meets generally accepted safe practices should be established for conducting these inspections and tests. The results of inspections, audits, monitoring, and tests, as well as corrective action, should be documented in writing, where appropriate.

In addition, all ANSI, NFPA, ASME, Compressed Gas Association, etc. standards that are incorporated by reference, and state OSHA standards must be consulted for these requirements. This list does not include those requirements.

Note: This checklist is current as of May 2003.

INDEX

Abrasive Wheels	1910.215(d)	Hazardous Waste Operations	1910.120
Acrylonitrile	1910.1045	Helicopters	1910.183
Air Receivers	1910.169	Hydrogen	1910.103
Ammonia	1910.111	Laboratories	1910.1450
Arsenic	1910.1018	Ladders	1910.25,.26,.27
Asbestos	1910.1001	Lead	1910.1025
Bakery	1910.263	Lockout/Tagout	1910.147
Benzene	1910.1028	Logging	1910.266
Bloodborne Pathogens	1910.1030	LP Gases	1910.110
Building Maintenance Platforms	1910.66	Manlifts	1910.68
Cadmium	1910.1027	Mechanical Power Transmission Apparatus	1910.219
Coke Ovens	1910.1029	Methylene Chloride	1910.1052
Compressed Gases	1910.101	Methylenedianiline	1910.1050
Confined Spaces	1910.146	Noise	1910.95
Cotton Dust	1910.1043	Oxygen	1910.104
Cranes	1910.179,.180	Personal Protective Equipment	1910.132(d)
DBCP	1910.1044	Power Presses	1910.217
Derricks	1910.181	Powered Platforms	1910.66,.67
Dip Tanks	1910.122-.126	Process Safety Management	1910.119
Diving	1910.421,.423,.430	Pulp, Paper, Paperboard	1910.261
Electrical	1910.304,.333-.335	Radiation	1910.1096
Electric Power Generation, Transmission, and Distribution	1910.269	Respirators	1910.134
Emergency Response	1910.120	Sawmills	1910.265
Ethylene Oxide	1910.1047	Scaffolds	1910.28
Exhaust Systems	1910.94	Slings	1910.184
Explosives	1910.109	Spray Finishing	1910.107
Fire Alarm Signaling Systems	1910.37,.165	Sprinklers	1910.159,.37
Fire Brigades	1910.156	Standpipe & Hose	1910.158
Fire Detection	1910.164	Telecommunications	1910.268
Fire Extinguishers	1910.157	Tools	1910.243,.244
Fixed Extinguishing	1910.160,.161, 162,.163	Vinyl Chloride	1910.1017
Flammable Liquids	1910.106	Welding	1910.252-.255
Forklifts	1910.178	Wheel Rims	1910.177

Figure A-10. OSHA requirements for inspections and audits task group. *(continued)*

CTJ Safety Associates	Inspections, Audits, Tests, and Monitoring Required by OSHA for General Industry (29 CFR 1910)		Date: May 2003
			Page 2 of 23

Standard	Subject	Coverage	Frequency	Comments
1910.25 (d)(1)(x)	Portable wood ladders	All wooden ladders	Frequently	
1910.26 (c)(2)(vi)	Portable metal ladders	Metal ladders that tip over, or are exposed to oil and grease	Immediately	
1910.27(f)	Fixed ladders	All fixed ladders	Regularly	
1910.28 (d)(14)	Scaffold frames	Welded frames and accessories	Periodic	
1910.28 (f)(11)	Masons multiple point suspension scaffolds	All points of the scaffold	Before installation; periodic	
1910.28 (g)(8)	Two-point suspension scaffold	Wire ropes, fiber ropes, slings, hangers, platforms, supports	Before installation; periodic	
1910.28 (i)(6)	Single-point suspension scaffold	Hoisting machines, cables, etc.	After installation; every 30 days	
1910.28 (p)(6)	Interior hung scaffold	Overhead supporting members	Before scaffold is erected	
1910.37 (m)	Automatic sprinkler systems	All systems	Periodic	
1910.37 (n)	Fire alarm signaling system	All systems tested (nonsupervised and supervised)	Nonsupervised— every 2 months; supervised— annually	
1910.66 (g)	Powered platforms for building maintenance	All powered platforms	Before being placed into service; every 12 months; before each use	
1910.67 (c)(2)(i)	Vehicle-mounted elevating and rotating work platforms	Extensible and articulating boom platform lift controls	Each day prior to use	
1910.67 (c)(2)(xii)	Vehicle-mounted elevating and rotating work platforms	Aerial lift before travel	Prior to moving	

Figure A-10. *(continued)*

CTJ Safety Associates	Inspections, Audits, Tests, and Monitoring Required by OSHA for General Industry (29 CFR 1910)		Date: May 2003
			Page 3 of 23

Standard	Subject	Coverage	Frequency	Comments
1910.67 (c)(3)	Vehicle-mounted elevating and rotating work platforms	Electrical tests	Per ANSI A92.2	
1910.68 (e)(1)	Manlifts	Steps, rails, landings, belt, illumination, motor, brakes, etc.	Every 30 days; limit switches checked weekly	
1910.94 (a)(3)(i)(e)	Abrasive blasting	Slit abrasion-resistant baffles	Regularly	
1910.94 (a)(4)(i)(b)	Abrasive blasting	Exhaust ducts static pressure	Upon completion of installation and periodically thereafter	
1910.94 (a)(6)	Abrasive blasting	Air supply and air compressors tested	Regularly	
1910.94 (b)(4)(ii)	Grinding, polishing, and buffing operations	Exhaust systems tested	Per ANSI Z9.2	
1910.94 (c)(3)(iii)(a)	Spray-finishing operations	Overspray filters	As needed	
1910.95 (d)	Noise	When employees may exceed 8-hour 85 dBA TWA	Initially; changes in production, process or equipment; new employees exposed; attenuation of hearing protectors inadequate	
1910.95 (h)(5)	Noise	Audiometer calibration	Daily, annual, every 2 years	
1910.101 (a)	Compressed gases	All compressed-gas cylinders	As stated by DOT and CGA	
1910.103 (b)	Gaseous hydrogen systems	Hydrogen containers and units	As stated by DOT, ASME, and CGA	
1910.103 (b)(1)(vi)	Gaseous hydrogen systems	All piping, tubing and fittings	After installation	

Figure A-10. *(continued)*

| CTJ Safety Associates | Inspections, Audits, Tests, and Monitoring Required by OSHA for General Industry (29 CFR 1910) | | Date: May 2003 |
| | | | Page 4 of 23 |

Standard	Subject	Coverage	Frequency	Comments
1910.103 (c)(1)(vii)	Liquefied hydrogen systems	All field-erected piping	After installation	
1910.104 (b)(4)(iii)	Bulk oxygen systems	High-pressure gaseous oxygen containers	As stated by ASME and DOT	
1910.104 (b)(8)(iv)	Bulk oxygen systems	All field-erected piping	After installation	
1910.106 (b)(2)(v)(i)	Aboveground tanks with flammable or combustible liquids	Flow capacity of tank venting devices 12 inches and smaller in nominal pipe size.	As appropriate	
1910.106 (b)(5)(vi) (L)	All tanks	Independent pumping units	Periodically	
1910.106 (b)(7)(i)	All tanks	Strength-tested	Before placement into service	
1910.106 (c)(7)	Piping, valves, and fittings	Hydrostatically or pneumatically tested; joints and connections	Before being covered, enclosed, or placed in use	
1910.106 (e)(5)(v)	Flammable and combustible liquids	All plant fire protection facilities	Periodically	
1910.106 (e)(8)	Flammable and combustible liquids	Hot work	Prior to the work	
1910.106 (f)(4)(vi)	Flammable and combustible liquids	Relief devices on pumps on wharves	Yearly	
1910.106 (f)(4)(vii)	Flammable and combustible liquids	Pressure hoses and couplings on wharves	Appropriate to the service	
1910.106 (g)(3)(v)(f)	Flammable and combustible liquids	Pressure piping system between pump discharge and connection for dispensing facility	After installation; every 5 years	
1910.106 (h)(6)(iv)	Flammable and combustible liquids	All plant fire protection facilities in processing plants	Periodically	

Figure A-10. *(continued)*

| CTJ Safety Associates | Inspections, Audits, Tests, and Monitoring Required by OSHA for General Industry (29 CFR 1910) | | Date: May 2003 |
| | | | Page 5 of 23 |
Standard	**Subject**	**Coverage**	**Frequency**	**Comments**
1910.106 (h)(7)(ii)(b)	Flammable and combustible liquids	Hot work	Prior to the work	
1910.107 (b)(5)(i)	Spray booths	Filter rolls	As needed	
1910.107 (e)(6)(iii)	Spray booths	Pressure hose and couplings	Appropriate to the service	
1910.109 (d)(2)(iii) (b)	Explosives	Fire extinguishers in vehicle used to transport explosives	Periodic	
1910.109 (d)(2)(iv)	Explosives	Motor vehicle used to transport explosives	As needed	
1910.109 (g)(2)(iv)(a)	Explosives	Sensitivity of the blasting agent	Regular intervals and after every formulation change	
1910.109 (h)(3)(v)(b)	Explosives (water gel)	Mixing, conveying, and electrical equipment	Daily	
1910.110 (b)(3)	Liquefied Petroleum (LP) gases	Container	As stated by ASME code	
1910.110 (b)(8)(ix)	LP gases	All piping tubing and hose	After assembly and installation	
1910.110 (b)(11)(ii) (c)	LP gases	Vaporizers of less than one quart capacity heated by ground or surrounding air	As needed	
1910.110 (g)(12)(i)	LP gases	DOT containers requalified	Per DOT regulation	
1910.110 (h)(9)(vii)	LP gases	Piping	After assembly	
1910.111 (d)(6)	Anhydrous ammonia	Containers for refrigerated storage that require field fabrication	After reassembly	
1910.119 (e)	Process hazard analysis	All processes of highly hazardous chemicals	Every 5 years	

Figure A-10. *(continued)*

CTJ Safety Associates	Inspections, Audits, Tests, and Monitoring Required by OSHA for General Industry (29 CFR 1910)		Date: May 2003
			Page 6 of 23

Standard	Subject	Coverage	Frequency	Comments
1910.119 (f)(3)	Processes of highly hazardous chemicals	Written operating procedures current	Annually	
1910.119 (h)(2)(v)	Processes of highly hazardous chemicals	Performance of contract employers	Periodically	
1910.119 (i)	Prestartup safety review of processes of highly hazardous chemicals	New and modified facilities	Prestartup	
1910.119 (j)(4)	Mechanical integrity of process equipment of highly hazardous chemicals	Inspections and tests of equipment	Manufacturer's recommendations and good engineering practices	
1910.119 (o)(1)	Processes of highly hazardous chemicals	Compliance audits	Every 3 years	
1910.120 (b)(4)(iv)	Hazardous-waste operations	Effectiveness of site safety–health plan	As needed	
1910.120 (c)	Hazardous-waste operations	All hazardous-waste sites	Preliminary evaluation	
1910.120 (h)	Hazardous-waste operations	Exposed employees monitoring	Initially; periodic	
1910.120 (j)(1)(iii) and (7)	Hazardous-waste operations	Drums and containers	Prior to being moved	
1910.120 (q)(3)(ii)	Emergency response	Emergency response to all conditions present	Initially	
1910.124 (e)	Dip tanks	Atmosphere tests before entering tanks	Prior to entry	
1910.124 (j)(1)	Dip tanks	Dip tank ventilation: hoods and ductwork	Quarterly during operations; after a prolonged shutdown	
1910.124 (j)(2)	Dip tanks	Airflow	Quarterly during operations; after a prolonged shutdown	

Figure A-10. *(continued)*

CTJ Safety Associates	Inspections, Audits, Tests, and Monitoring Required by OSHA for General Industry (29 CFR 1910)		Date: May 2003
			Page 7 of 23
Standard	**Subject**	**Coverage**	**Frequency**	**Comments**
1910.124 (j)(3)	Dip tanks	Dipping and coating equipment	Periodic	
1910.132 (d)	PPE (eye, face, head, foot, hand)	Each workplace	Initially	
1910.133	PPE (eye and face)	Each workplace	Initially	
1910.134 (d)(1)(iii)	Respirators	Work area conditions and degree of employee exposure	Prior to respirator selection and as needed	
1910.134 (f)(1)–(3)	Respirators	Qualitative and quantitative fit test for tight-fitting facepiece respirator	Initially; change in facepiece or physical conditions; annually	
1910.134 (g)(2)(i)	Respirators	Work area conditions and employee exposure or stress	Continually	
1910.134 (h)(3)(i)(A)	Respirators	All respirators in routine use	Before each use; during cleaning	
1910.134 (h)(3)(i)(B)	Respirators	All respirators for emergency use (for proper functioning)	Before and after each use; monthly; as stated by manufacturer's recommendations	
1910.134 (h)(3)(i)(C)	Respirators	All respirators for emergency escape-only use	Before being carried into the workplace for use	
1910.134 (h)(3)(iii)	Respirators	All SCBAs	Monthly	
1910.134 (h)(3)(iii)	Respirators	Air and oxygen cylinders	Recharge when pressure falls to 90% of recommended Pressure level.	
1910.134 (i)(4)(i)	Respirators	Breathing air cylinders	As stated by DOT	
1910.134 (l)	Respirators	Effectiveness of the program	Regularly	
1910.135	PPE (head)	Each workplace	Initially	

Figure A-10. *(continued)*

CTJ Safety Associates	Inspections, Audits, Tests, and Monitoring Required by OSHA for General Industry (29 CFR 1910)		Date: May 2003
			Page 8 of 23

Standard	Subject	Coverage	Frequency	Comments
1910.136	PPE (foot)	Each workplace	Initially	
1910.137 (b)(2)(ii)	Electrical protective equipment	All insulating equipment	Before each day's use/following any incident	
1910.137 (b)(2)(viii)	Electrical protective equipment	All insulating equipment	Periodic	
1910.138	PPE (hand)	Each workplace	Initially	
1910.146 (c)(1)	Confined spaces	Identify permit spaces	Initially/when conditions change	
1910.146 (c)(5)(ii)(C)	Confined spaces	Internal atmosphere tested	Prior to entry	
1910.146 (c)(5)(ii)(F)	Confined spaces	Internal atmosphere retested	Periodically	
1910.146 (c)(5)(ii)(G)	Confined spaces	Atmospheric testing of permit spaces upon detection of a hazardous atmosphere	Upon detection of hazardous atmosphere	
1910.146 (c)(5)(ii)(H)	Confined spaces	Employer verifies space is safe for entry and preentry measures have been taken	Prior to entry	
1910.146 (d)(5)(v)	Confined spaces	Space is reevaluated if entrant/representative has reason to believe first evaluation is inadequate	As needed	
1910.146 (k)(1)(i)	Confined spaces	Prospective rescuer's ability to respond to a rescue in a timely manner	Prior to selection of a rescuer	
1910.146 (k)(1)(ii)	Confined spaces	Prospective rescuer's ability in rescuing entrants from particular spaces identified	Prior to selection of a rescuer	
1910.147 (c)(6)	Lockout/tagout	Energy control procedures are implemented	At least annually	

Figure A-10. *(continued)*

CTJ Safety Associates	Inspections, Audits, Tests, and Monitoring Required by OSHA for General Industry (29 CFR 1910)		Date: May 2003
			Page 9 of 23
Standard	**Subject**	**Coverage**	**Frequency**	**Comments**
1910.147 (d)(6)	Lockout/tagout	Verification that energy is isolated and deenergized	Prior to starting work	
1910.147 (e)(1), (2)	Lockout/tagout	Ensure components are operationally intact, employees safely positioned	Prior to removing lock or tag	
1910.156 (d)	Fire brigades	Firefighting equipment	Annually; fire extinguishers and respirators monthly	
1910.157 (e)	Portable fire extinguishers	Portable extinguishers or hoses used in lieu of extingnishers	Visual monthly; maintenance annual; dry chemical 6 years	
1910.157(f) (2)–(12)	Portable fire extinguishers	Hydrostatic checks of cylinders and hose assembly	As stated in table L-1; when in evidence of corrosion or injury	
1910.158 (e)(1)(i)	Standpipe–hose systems	Hydrostatic checks of class II and III systems	After installation	
1910.158 (e)(2)(iii)	Standpipe–hose systems	Hose systems	Annually; after each use	
1910.158 (e)(2)(v)	Standpipe–hose systems	Hemp or linen hose	Annually	
1910.159 (c)(2)	Automatic sprinkler systems	Main drain flow, test valve	Annually; every 2 years	
1910.159 (c)(3)	Automatic sprinkler systems	Acceptance tests	After installation	
1910.159 (c)(8)(ii)	Automatic sprinkler systems	When using older style sprinklers to replace standard sprinklers	Prior to installation	
1910.160 (b) (6)-(9)	Fixed extinguishing systems—fixed	Systems, containers	Systems—annual; containers—semiannual	
1910.161 (a)	Fixed extinguishing systems—drychemical	All such systems and containers	Same as 1910.160	

Figure A-10. *(continued)*

CTJ Safety Associates	Inspections, Audits, Tests, and Monitoring Required by OSHA for General Industry (29 CFR 1910)		Date: May 2003
			Page 10 of 23
Standard	**Subject**	**Coverage**	**Frequency**	**Comments**
1910.161 (b)(4)	Fixed extinguishing systems—dry chemical	Dry chemical supply	Annually	
1910.162 (a)(1)	Fixed extinguishing systems—gaseous	All such systems and containers	Same as 1910.160	
1910.163 (a)(1)	Fixed extinguishing systems—water spray and foam	All such systems and containers	Same as 1910.160	
1910.164 (c)(2)	Fire detection systems	Fire detectors and detection systems	As often as needed	
1910.165 (d)(2)	Employee alarm systems	Nonsupervised alarm systems	Every 2 months	
1910.165 (d)(4)	Employee alarm systems	Supervised alarm systems	Annual	
1910.169 (b)(3)(iv)	Air receivers	Safety valves	Frequently and regularly	
1910.177 (d)(3)(iii)	Multipiece and single-piece rim wheels	Restraining devices and barriers	Prior to each day's use and after any rim separation	
1910.177 (e)(2)	Multipiece and single-piece rim wheels	Components of rim wheels	Prior to assembly	
1910.177 (f)(7)	Multipiece and single-piece rim wheels	Tire and wheel components	After inflation	
1910.178 (m)(7)	Powered industrial trucks	Flooring of trucks, trailers and railroad cars	Before being driven into	
1910.178 (q)(7)	Powered industrial trucks	All trucks	Beginning of each shift	
1910.179 (b)(3)	Overhead and gantry cranes	Modified cranes rerating	Upon modification	
1910.179 (j)(1)(i)	Overhead and gantry cranes	New and altered cranes	Prior to use	
1910.179 (j)(1)(ii)	Overhead and gantry cranes	Cranes in regular service	Daily, monthly, annually	
1910.179 (k)	Overhead and gantry cranes	New and altered cranes testing	Prior to initial use	

Figure A-10. (continued)

CTJ Safety Associates	Inspections, Audits, Tests, and Monitoring Required by OSHA for General Industry (29 CFR 1910)		Date: May 2003
			Page 11 of 23
Standard	Subject	Coverage	Frequency	Comments
1910.179 (m)	Overhead and gantry cranes	Ropes	Monthly	
1910.179 (n)(3)(vii)	Overhead and gantry cranes	Brakes	Each time the load approaches the rated load	
1910.179 (n)(4)	Overhead and gantry cranes	Hoist limit switch	Beginning of each shift	
1910.180 (d)	Crawler, locomotive, and truck cranes	New and altered cranes	Prior to initial use	
1910.180 (e)	Crawler, locomotive, and truck cranes	Each new production crane	Final assembly	
1910.180 (g)	Crawler, locomotive, and truck cranes	Ropes	Monthly	
1910.180 (h)(3)(viii)	Crawler, locomotive, and truck cranes	Brakes	Each time the load approaches the rated capacity	
1910.181 (d)	Derricks	New and altered derricks	Prior to use, daily, monthly, annually	
1910.181 (e)	Derricks	New and altered derricks testing	Prior to initial use	
1910.181 (g)	Derricks	Ropes	Monthly	
1910.181 (i)(3)(vii)	Derricks	Brakes	Each time the load approaches the rated capacity	
1910.183 (d)	Helicopters	Cargo hooks tested	Prior to each day's use	
1910.184 (d)	Slings	Sling, all fastenings and attachments	Prior to each day's use; during sling use	
1910.184 (e)(3)	Slings	Alloy steel chain slings in use	At least annually	
1910.184 (e)(4)	Slings	Each new, repaired or reconditioned alloy steel chain sling proof test	Prior to use	

Figure A-10. (continued)

CTJ Safety Associates	Inspections, Audits, Tests, and Monitoring Required by OSHA for General Industry (29 CFR 1910)		Date: May 2003
			Page 12 of 23
Standard	**Subject**	**Coverage**	**Frequency**	**Comments**
1910.184 (e)(7)	Slings	Repairing and reconditioning alloy steel chain sling when welding or heat testing is performed	Before use	
1910.184 (f)(4)(ii)	Slings	Wire rope slings end attachment proof test	Prior to initial use	
1910.184 (g)(5)	Slings	Metal mesh slings proof test	Before use	
1910.184 (i)(8)(ii)	Slings	Synthetic web slings repaired proof test	Prior to its return to service	
1910.215 (d)	Abrasive Wheel machinery	All wheels, spindle speed, ring test	Immediately before mounting	
1910.217 (c)(3)(iv)(d)	Mechanical power presses	Each pullout device in use	Start of each shift, after die set-up, changing operators	
1910.217 (e)	Mechanical power presses	All presses	Weekly	
1910.217 (h)(10)	Presence sensing device initiation of mechanical power presses	PSDI, bearings, automatic lubrication systems, clutch and brake mechanisms	Beginning of each shift, whenever a die change is made	
1910.218 (a)(2)	Forging machines	Forge shop equipment	Regular	
1910.219 (p)(1)	Mechanical power transmission apparatus	All equipment	Every 60 days	
1910.219 (p)(4)	Mechanical power transmission apparatus	Hangers	As needed	
1910.219 (p)(6)(ii)	Mechanical power transmission apparatus	Belts, lacings, and fasteners	As needed	
1910.243 (c)(5)	Portable powered tools	Portable abrasive wheels, spindle speed	Immediately before mounting	

Figure A-10. *(continued)*

CTJ Safety Associates	Inspections, Audits, Tests, and Monitoring Required by OSHA for General Industry (29 CFR 1910)		Date: May 2003
			Page 13 of 23

Standard	Subject	Coverage	Frequency	Comments
1910.243 (d)(4)	Portable powered tools	Explosive actuated fastening tools	Before using and regularly	
1910.244 (a)(2)(vi)	Portable equipment	Jacks	No less than every 6 months	
1910.252 (a)(2)(iv)	Welding: general	All operations of welding and cutting	Before beginning	
1910.252 (b)(3)	Welding: general	Used drums, barrels, tanks, other containers	Before any hot work	
1910.252 (d)(1)(vii)	Welding: general	X- rays and radioactive isotopes for transmission pipelines	As stated in ANSI Z54.1	
1910.252 (d)(2)(ii)	Welding: general	X- rays and radioactive isotypes for mechanical piping systems	As stated in ANSI Z54.1	
1910.253 (a)(4)	Welding and cutting: oxygen– fuel gas	Oxygen–fuel gas systems, workers' competency	Before being left in charge	
1910.253 (c)(3)(iii)	Welding and cutting: oxygen– fuel gas	Assembled low- pressure oxygen manifold	Before use	
1910.253 (d)(3)(vii)	Welding and cutting: oxygen– fuel gas	Fittings and pipes on oxygen–fuel gas systems	Before assembly	
1910.253 (d)(5)(i)	Welding and cutting: oxygen– fuel gas	Oxygen–fuel gas piping systems	Before being placed into service	
1910.253 (e)(6)(iv)	Welding and cutting: oxygen– fuel gas	Oxygen–fuel gas union nuts and connections on regulators	Before use	
1910.253 (f)(2)(ii)	Welding and cutting: oxygen– fuel gas	Acetylene generator relief valves	Regularly	
1910.253 (g)(3)(ii)	Welding and cutting: oxygen– fuel gas	Calcium carbide storage containers	Periodic	

Figure A-10. *(continued)*

CTJ Safety Associates	Inspections, Audits, Tests, and Monitoring Required by OSHA for General Industry (29 CFR 1910)		Date: May 2003
			Page 14 of 23

Standard	Subject	Coverage	Frequency	Comments
1910.254 (c)(2)(v)	Welding and cutting: arc welding and cutting	Ground connections checked	Before use	
1910.254 (d)(2)&(3)	Welding and cutting: arc welding and cutting	Arc welding, machine hookup	Before starting operation	
1910.254 (d)(9)(ii))	Welding and cutting: arc welding and cutting	Arc welders that have become wet	Before use	
1910.255 (e)	Welding: resistance	Resistance welding machines	Periodic	
1910.261 (b)(5)	Pulp, paper, paperboard mills	Vessel entering	Prior to entry	
1910.261 (g)(3)	Pulp, paper, paperboard mills	Acid tower structure during winter months	Daily	
1910.261 (g)(4)	Pulp, paper, paperboard mills	Acid tanks cleanliness	Before entry	
1910.261 (g)(7)	Pulp, paper, paperboard mills	Acid storage tank hoops	Periodic	
1910.261 (g)(10)	Pulp, paper, paperboard mills	Gas masks in digester building	Monthly	
1910.261 (g)(15)(iii)	Pulp, paper, paperboard mills	Ladders and lifelines at digester	Before each use	
1910.261 (g)(16)(ii)	Pulp, paper, paperboard mills	Pressure tanks—accumulators	Every 6 months	
1910.261 (g)(17)	Pulp, paper, paperboard mills	Pressure vessel safety valves	Between each cook	
1910.261 (g)(18)(iii)	Pulp, paper, paperboard mills	Heavy duty pipe, valves, fittings between digester and blow pit	Every 6 months	
1910.261 (h)(3)(ii)	Pulp, paper, paperboard mills	Bleaching gas masks	Regularly	
1910.261 (k)(1)	Pulp, paper, paperboard mills	Machine room emergency stops	Periodically	
1910.261 (k)(24)(ii)	Pulp, paper, paperboard mills	Lifting equipment and reels	Regularly	

Figure A-10. *(continued)*

CTJ Safety Associates	Inspections, Audits, Tests, and Monitoring Required by OSHA for General Industry (29 CFR 1910)		Date: May 2003
			Page 15 of 23

Standard	Subject	Coverage	Frequency	Comments
1910.261 (l)(2)	Pulp, paper, paperboard mills	Finishing room emergency stops	When stopping the machine	
1910.263 (h)(2)	Bakery equipment	Emergency stop bar on manually fed dough brakes	Every 30 days	
1910.263 (l)(3)(ii)	Bakery equipment	Piping at ovens	As needed	
1910.263 (l)(9)(ii)	Bakery equipment	Safety devices on ovens	Twice per month by plant—annually by manufacturer	
1910.263 (l)(9)(v)(b)	Bakery equipment	Oven safety shutoff valve	Twice per month	
1910.263 (l)(15)(iii)	Bakery equipment	Oven duct systems	Initially, every 6 months	
1910.265 (c)(12)(vi)	Sawmills	Concealed conductors	Prior to any work	
1910.265 (c)(19)	Sawmills	Foundations of stationary tramways and trestles	Frequently	
1910.265 (c)(22)	Sawmills	Mechanical power-transmission apparatus	Every 60 days	
1910.265 (c)(24)(iv)	Sawmills	Slings	Daily	
1910.265 (c)(24)(v)	Sawmills	Ropes and cables	Weekly	
1910.265 (c)(24)(viii)(c)	Sawmills	Wire rope clips attached with U bolts	Frequently	
1910.265 (c)(24)(ix)(a)	Sawmills	Chains	Before initial use/weekly	
1910.265 (c)(26)(ix)	Sawmills	Stacker and unstacker	Frequently	
1910.265 (c)(30)(x)	Sawmills	Lift trucks	See 1910.178	
1910.265 (d)(2)(iii)(d)	Sawmills	Log handling dogging lines	Periodic	

Figure A-10. *(continued)*

CTJ Safety Associates	Inspections, Audits, Tests, and Monitoring Required by OSHA for General Industry (29 CFR 1910)		Date: May 2003
			Page 16 of 23
Standard	**Subject**	**Coverage**	**Frequency**	**Comments**
1910.265 (d)(2)(iv) (d)	Sawmills	Log handling bilge area	As needed	
1910.265 (e)(2)(i)(b)	Sawmills	Band head saws	As needed	
1910.265 (e)(2)(ii)(b)	Sawmills	Band head saw wheels	Monthly	
1910.266 (d)(1)(ii)	Logging	Personal protective equipment	Before initial use on each shift	
1910.266 (d)(2)(iii)	Logging	First aid kits	Annually	
1910.266 (d)(6)(iv)	Logging	Each employee accounted for	End of each shift	
1910.266 (e)(1)(ii)	Logging	Tools	Before use on each shift	
1910.266 (f)(1)(ii)	Logging	Machinery	Before use on each shift	
1910.266 (g)(2)	Logging	Vehicles	Before use on each shift	
1910.266 (h)(1)(vii)	Logging	Danger trees	Before felling or removal	
1910.266 (h)(2)(ii) and (iii)	Logging	Each tree	Before felling	
1910.268 (b)(3)	Telecommunications	First aid kits	Monthly	
1910.268 (b)(6)	Telecommunications	Support structures	Prior to employee being on structure	
1910.268 (e)	Telecommunication	Tools and protective equipment	Before each day's use	
1910.268 (f)(5)	Telecommunications	Insulating gloves, blankets, etc.	See 1910.268(f)(5)	
1910.268 (f)(9)	Telecommunications	Rubber gloves	Prior to each day's use	
1910.268 (g)(1)	Telecommunications	Personal climbing equipment	Prior to each day's use	

Figure A-10. *(continued)*

CTJ Safety Associates	Inspections, Audits, Tests, and Monitoring Required by OSHA for General Industry (29 CFR 1910)		Date: May 2003
			Page 17 of 23

Standard	Subject	Coverage	Frequency	Comments
1910.268 (g)(2) (ii)(A)(2)	Telecommunications	All fabric and leather	As needed	
1910.268 (g)(3)(iii)	Telecommunications	Pole climbers	Before each day's use/weekly	
1910.268 (h)(1)	Telecommunications	Ladders	Before anyone is on ladder	
1910.268 (j)(1)	Telecommunications	Vehicle-mounted materials handling devices	Daily at beginning of shift	
1910.268 (j)(4)(iv)(f)	Telecommunications	Derrick trucks	At least annual per manufacturer's specs	
1910.268 (k)(1)	Telecommunications	Poles	Prior to unloading	
1910.268 (m)(3)	Telecommunications	Vertical power conduit, power ground wires, street light fixtures tested for voltage	As needed	
1910.268 (m)(7)(iv)	Telecommunications	Radiofrequency line wires	Before handling	
1910.268 (n)(2)	Telecommunications	Wood poles	Before climbing	
1910.268 (n)(5), (6)	Telecommunications	Cable suspension strand	Before attaching a splicing platform	
1910.268 (o)(2), (5)(i)	Telecommunications	Manholes	Prior to entering	
1910.268 (q)(2)	Telecommunications	Tree trimming	Before climbing, entering, or working around any tree	
1910.269 (a)(2)(iii)	Electric power generation, transmission, and distribution	Employees following proper work practices	Annually	

Figure A-10. *(continued)*

CTJ Safety Associates	Inspections, Audits, Tests, and Monitoring Required by OSHA for General Industry (29 CFR 1910)		Date: May 2003
			Page 18 of 23
Standard	**Subject**	**Coverage**	**Frequency**	**Comments**
1910.269 (a)(3)	Electric power generation, transmission, and distribution	Existing conditions on or near electric lines and equipment	Before work is started	
1910.269 (b)(3)	Electric power generation, transmission, and distribution	First aid kits	Frequently; annually	
1910.269 (d)(2)(v)	Electric power generation, transmission, and distribution	Energy control procedure	Annually	
1910.269 (d)(6)(vii)	Electric power generation, transmission, and distribution	Verification of energy control	Before starting work	
1910.269 (e)(4)	Electric power generation, transmission, and distribution	Enclosed-space hazards	Before entry cover is removed	
1910.269 (e)(8)	Electric power generation, transmission, and distribution	Calibration of test instruments	Per manufacturer's specifications	
1910.269 (e)(9)	Electric power generation, transmission, and distribution	Test for oxygen deficiency in enclosed spaces	Before entry	
1910.269 (e)(10)	Electric power generation, transmission, and distribution	Test for flammable gases and vapors in enclosed spaces	Before entry	
1910.269 (e)(14)	Electric power generation, transmission, and distribution	Test for flammable gases and vapors if open flames used in enclosed spaces	Once per hour or More frequently	
1910.269 (f)	Electric power generation, transmission, and distribution	Excavation	Daily	
1910.269 (g)(2)(ii)	Electric power generation, transmission, and distribution	Body belts, lanyards, lifelines, harnesses	Daily before use	

Figure A-10. *(continued)*

CTJ Safety Associates	Inspections, Audits, Tests, and Monitoring Required by OSHA for General Industry (29 CFR 1910)		Date: May 2003
			Page 19 of 23

Standard	Subject	Coverage	Frequency	Comments
1910.269 (j)(2)(i)	Electric power generation, transmission, and distribution	Live-line tools	Daily before use	
1910.269 (l)(9)	Electric power generation, transmission, and distribution	Non-current-carrying metal parts	Before exposures	
1910.269 (n)(5)	Electric power generation, transmission, and distribution	Lines and equipment	Before grounding	
1910.269 (o)(2)	Electric power generation, transmission, and distribution	Field test area safety checks	Periodic	
1910.269 (o)(6)	Electric power generation, transmission, and distribution	Safety practices at field test areas	At the beginning of each series of tests	
1910.269 (p)(l)(i)	Electric power generation, transmission, and distribution	Mechanical elevating and rotating equipment	Before use on each shift	
1910.269 (q)(l)(i)	Electric power generation, transmission, and distribution	Elevated structures subject to stresses such as climbing or installation of equipment	Before applying stresses	
1910.269 (q)(2)(iv)	Electric power generation, transmission, and distribution	Lines installed parallel to existing energized lines	Before lines are installed	
1910.269 (q)(3)(x)	Electric power generation, transmission, and distribution	All bucket controls in working condition	Before employee elevation	
1910.269 (q)(3)(xii)	Electric power generation, transmission, and distribution	Boom-current test for live-line bare-hand work	Before work each day	

Figure A-10. *(continued)*

CTJ Safety Associates	Inspections, Audits, Tests, and Monitoring Required by OSHA for General Industry (29 CFR 1910)		Date: May 2003
			Page 20 of 23

Standard	Subject	Coverage	Frequency	Comments
1910.269 (r)(1)(i)	Electric power generation, transmission, and distribution	Determine nominal voltage for line clearance tree trimming operations	Before climbing or working around tree	
1910.269 (r)(7)(ii)	Electric power generation, transmission, and distribution	Climbing ropes	Before each use	
1910.269 (t)(6)	Electric power generation, transmission, and distribution	Energized cables to be moved in underground installations	Before moving	
1910.269 (v)(6)(i)	Electric power generation, transmission, and distribution	Power generation and steam spaces	Before work is permitted and after its completion	
1910.269 (v)(9)(i)	Electric power generation, transmission, and distribution	Overhead areas at boilers in power generation	Before work is started	
1910.269 (w)(5)(ii)	Electric power generation, transmission, and distribution	Personal flotation devices	Frequently	
1910.272 (g)	Grain handling facilities	Entry into bin silos, tanks	Prior to entry	
1910.272 (m)(1)(i)	Grain handling facilities	Control equipment for dryers, grain stream processing, dust collection, bucket elevators	Regular and per manufacturer	
1910.303 (b)(1)(ii)	Wiring design	Assured equipment grounding program cord set, cord- and plug-connected equipment	Before each day's use	
1910.304 (f)(7)(ii)(c)	Wiring design	Grounding continuity of equipment 1000 V and over	Continuous	

Figure A-10. *(continued)*

CTJ Safety Associates	Inspections, Audits, Tests, and Monitoring Required by OSHA for General Industry (29 CFR 1910)		Date: May 2003
			Page 21 of 23
Standard	Subject	Coverage	Frequency	Comments
1910.333 (b)(2)(iv) (B)	Selection and use of electrical safety work practices	Electrical parts to which employees may be exposed tested	Before servicing and maintenance	
1910.333 (b)(2)(v)(a)	Selection and use of electrical safety work practices	Reenergizing equipment	Before reenergizing	
1910.334 (a)(2)	Use of electrical equipment	Portable cord- and plug-connected equipment and extension cords	Each shift	
910.334 (c)(2)	Use of electrical equipment	Test instruments and equipment	Before use	
1910.335 (a)(1)(ii)	Electrical safeguards for personnel protection	Protective equipment	Periodically	
1910.421 (g)	Diving	Breathing gas supply system	Prior to each dive	
1910.423 (e)(1)	Diving	Decompression sickness	Within 45 days	
1910.430 (b)(4)	Diving	Output of air compressor systems	Every 6 months	
1910.430 (c)(1)(iii)	Diving	Breathing gas supply hoses	Annually	
1910.430 (g)(2)	Diving	Depth gauges	Every 6 months	
1910.1001 (d)	Asbestos	Where exposure occurs	Initially; periodic	
1910.1001 (g)(4)(ii)	Asbestos	Respirator fit tests	Initially; every 6 months	
1910.1001 (j)(8)(ii)(A)	Asbestos	To rebut the designation of PACM	Initially	
1910.1017 (d)	Vinyl chloride	Exposure monitoring	Initially; monthly, quarterly, process change	
1910.1018 (e)	Inorganic arsenic	Where exposure occurs	Initially; quarterly, every 6 months, process change	

Figure A-10. *(continued)*

CTJ Safety Associates	Inspections, Audits, Tests, and Monitoring Required by OSHA for General Industry (29 CFR 1910)		Date: May 2003
			Page 22 of 23
Standard	**Subject**	**Coverage**	**Frequency**	**Comments**
1910.1018 (k)(5)	Inorganic arsenic	Dust collection and ventilation equipment	Periodic	
1910.1025 (d)	Lead	Exposure monitoring	Initially; periodic	
1910.1025 (e)(5)	Lead	Ventilation system	Every 3 months; within 5 days of any change	
1910.1027 (d)	Cadmium	Where exposure occurs	Initially; every 6 months; changes	
1910.1027 (f)(3)	Cadmium	Mechanical ventilation— measurement of effectiveness	as necessary; within 5 days of changes	
1910.1028 (e)	Benzene	Exposure monitoring	Initially; periodic, change in process after a spill/leak occurs	
1910.1029 (e)	Coke oven emissions	Where exposure occurs	Initially/periodic	
1910.1029 (f)	Coke oven emissions	Work practice controls	Prior to each charge	
1910.1030 (c)(2)	Bloodborne pathogens	Exposure determination	Upon employee exposure	
1910.1030 (d)(2)(ii)	Bloodborne pathogens	Engineering controls	Regular schedule	
1910.1030 (d)(2)(xiv)	Bloodborne pathogens	Contaminated equipment	Prior to servicing or shipping	
1910.1030 (d)(3)(ii)	Bloodborne Pathogens	Personal protective equipment use declined	Afterward	
1910.1030 (d)(3)(ix) (D)(i)	Bloodborne pathogens	No routine gloving for phlebotomies	Periodically	
1910.1030 (d)(4)(ii)(C)	Bloodborne pathogens	Receptacles intended for reuse	Regularly	

Figure A-10. *(continued)*

CTJ Safety Associates	Inspections, Audits, Tests, and Monitoring Required by OSHA for General Industry (29 CFR 1910)		Date: May 2003
			Page 23 of 23
Standard	**Subject**	**Coverage**	**Frequency**	**Comments**
1910.1030 (e)(4)(vi)	HIV and HBV research laboratories and production facilities	Ducted exhaust-air ventilation system	Initially	
1910.1043 (d)	Cotton dust	Where exposure occurs	Initially; periodic	
1910.1043 (e)(4)	Cotton dust	Where exposure occurs	Reasonable intervals	
1910.1044 (f)	DBCP	Exposure monitoring	Initially; periodic; process change	
1910.1045 (e)	Acrylonitrile	Where exposure occurs	Initially; periodic; process change	
1910.1047 (d)	Ethylene oxide	Where exposure occurs	Initially; periodic	
1910.1048 (d)	Formaldehyde	Where exposure occurs	Initially; periodic	
1910.1048 (j)	Formaldehyde	Leak and spill detection	Regular	
1910.1050 (e)	Methylenedianiline	Where exposure occurs	Initially; periodic	
1910.1050 (e)(8)	Methylenedianiline	Potential exposure of face, hands, forearms	Routine	
1910.1050 (l)	Methylenedianiline	Leak and spill detection	Regular	
1910.1052 (d)	Methylene chloride	Exposure monitoring	Initially; periodic; change in workplace conditions; spill/leak	
1910.1096 (d)	Ionizing radiation	Survey of materials and equipment and levels of radiation	As necessary	
1910.1096 (f)(3)	Ionizing radiation	Signal generating system	Initially; periodic	
1910.1200 (d)	Hazard communication	Chemical manufacturers; importers	Initially	
1910.1450 (d)	Laboratories	Where exposures may exceed AL or PEL	Initially; periodic	

Figure A-10. *(continued)*

HOUSEKEEPING RATING FORM						Page 1 of 2	
Area:		**Date inspected:**					
Inspector:							
Instructions: Circle the appropriate score under the Item Ratings opposite the item being evaluated. Place circled score in the score column. Add ratings for your total score. Record in overall rating block at the top of form and check overall category.							

			Item Ratings				
Machinery and Equipment	No Credit 0	Very Poor 0–18	Poor 18–35	Fair 35–52	Good 52–74	Very Good 74–100	**Score**
1. Must be clean and free of unnecessary material	0	.5	1	1.5	2	3	
2. Must be free of unnecessary oil or grease.	0	1	2	3	4	5	
3. Must have proper guards provided and in good condition.	0	1.5	2.5	3.5	5	7	
Stock and Material 1. Must be properly piled, secured, and arranged.	0	1.5	3	4.5	6	8	
2. Must be loaded safely and orderly in pans, cars, carts and trucks.	0	1.5	2.5	3.5	5	7	
Tools 1. Must be properly stored in assigned locations.	0	1	2	3	4.5	6	
2. Must be free of oil and grease.	0	.5	1	1.5	2	3	
3. Must be in safe working condition.	0	1	2	3	4.5	6	
Aisles 1. Must be provided for access to work positions, fire extinguishers, fire blankets, electrical disconnects, and safety showers.	0	1	2	3	4.5	6	
2. Must be safe and free of obstruction.	0	1	2	3	4.5	6	
3. Must be clearly marked.	0	.5	1	1.5	2	3	

Figure A-11. Sample housekeeping rating form. *(continued)*

Machinery and Equipment	No Credit 0	Very Poor 0–18	Poor 18–35	Fair 35–52	Good 52–74	Very Good 74–100	Score
Item Ratings							
Floors							
1. Must have surfaces safe, not overloaded, and suitable for work.	0	1	2	3	4.5	6	
2. Must be clean, dry and free of waste, unnecessary material, oil and grease.	0	1	2	3	4.5	6	
3. Must have an adequate number of waste receptacles provided.	0	.5	1	1.5	2	3	
Buildings							
1. Must have walls and windows that are properly maintained and reasonably clean for operations, and free of unnecessary items.	0	.5	1	1.5	2	3	
2. Must have lighting systems that are maintained in a clean and efficient manner.	0	.5	1	1.5	2	3	
3. Must have safe stairs that are clean, free of materials, well lighted, provided with adequate handrails and treads in good condition.	0	1	2	3	4	5	
4. Must have platforms that are guarded, clean, free of materials, and well lighted.	0	.5	1	2	3	4	
Grounds							
1. Must be in good order, free of refuse and unnecessary materials.	0	2	4	6	8	10	
Total Score:	0	18	35	52.5	74	100	

Note: Record the reason for all scores below "very good" on the back of this form.

Figure A-11. *(continued)*

	Housekeeping Rating Form	Page 2 of 2
Area:	**Date Inspected:**	**Inspector(s):**

Explanation for scores below "very good".

Machinery and equipment

Stock and material

Tools

Aisle

Floors

Buildings

Grounds

Other comments

Route copies to (1) work area; (2) Housekeeping Task Group; (3) safety coordinator; (4) file

Figure A-11. *(continued)*

	REQUEST FOR HOUSEKEEPING ASSISTANCE	Date:
		Page 1 of 1
Organization:	Facility:	Area/unit:

Brief description of housekeeping problem:

Basic cause(s) of problem:

Possible solutions to problem:

Specific Housekeeping Task Group assistance needed:

Housekeeping Task Group member assigned to follow-up:	Date assigned:	Target completion date:
Area/unit manager:		Date:
Housekeeping Task Group chairperson:		Date received:

Figure A-12. Sample form for requesting housekeeping assistance.

Appendix B
Frequently Asked Questions

Q1. How much time is required for task group members to effectively complete assignments?

A1. First, it should be recognized that assignments are never really completed; assignments are maintained. In other words, as long as there are employees, there will be a need for employee training, inspections, audits, updating policies, and so on. Task group members should devote only as much time as can be spared from their full-time jobs. Most assignments require only a few minutes each week or month to keep up. Some may require more time, but it is usually easy to find a balance whereby task group members can perform their jobs and still maintain assignments effectively.

Q2. What if no one volunteers to serve on task groups?

A2. It is rare when there are insufficient volunteers willing to serve on task groups. Most employees recognize EHS as a

Effective Environmental, Health, and Safety Management Using the Team Approach, by Bill Taylor
Copyright © 2005 John Wiley & Sons, Inc.

responsibility and are willing to get involved in management of the system. On the rare occasion where there are not enough volunteers, one approach is to name co-chairs with an hourly employee serving as a co-chair of each task group along with a department manager. This may motivate others to serve as volunteer members.

Q3.   What do we do with task group members who do not attend meetings or complete assignments?

A3.   It is frustrating and counterproductive to have task group members who do not participate. Task group members who consistently miss meetings or fail to complete assignments should eventually be replaced. The EHS manager should attend each task group meeting and should focus energy on those who participate rather than those who don't.

Q4.   Should task group members get a new assignment each month?

A4.   No. Assignments are typically things that are required as a part of successful EHS management. They are things that occur on an ongoing basis that must be done by someone, such as ensuring that employees receive required training. The purpose in giving the assignments is twofold: (1), it gets employees directly involved in the EHS process, thus increasing the awareness level of each participating employee; and (2), it is a way to ensure that the multitude of things that are required to be done, get done. Assignments are intended to be permanent. Assignments given to task group members should not change, but instead are to be passed on to upcoming task group members.

Q5.   How long does it usually take for task group members to complete their assignments?

A5.   As stated in the previous answer, EHS is a never-ending process. Task group members don't complete assignments, they maintain them. See A4.

Q6.   Should task group members attend FSHC meetings?

A6.   By all means. It is suggested that each task group member attend at least one meeting of the FSHC because there might be a day when he/she has to attend either as a representative of his/her task group or to give a report or answer questions related to his/her assignment. It is helpful to have an idea of how the FSHC meetings function and what to expect beforehand.

Q7.   Is it feasible for a task group to complete all that it is responsible for by meeting monthly?

A7.   We must keep in mind that EHS is a never-ending process. Just as a task group member does not finish his/her assignment, a task group never completes everything that it is responsible for. Since there will always be a need for audits, inspections, training and rules, and so forth, there will always be something for task groups to do. Is it feasible for the task group to keep up with its responsibilities with a monthly meeting? Absolutely.

Q8.   What if my task group wants to meet more than once per month?

A8.   That is not an unusual request by some task groups and task group members. But task groups should resist the urge to do this. It must be remembered that task group members all have full-time jobs and to spend more time than is necessary in task group meetings will create more problems than it will solve. Task groups should simply take a slow steady pace and get the job done. Since there is no end to the given areas of responsibilities there should be no urgency to get things completed.

Q9.   Who should keep minutes for the FSHC meetings?

A9.   There are no hard-and-fast rules regarding who should do this job. Normally the facility EHS manager would do this.

Q10.   What should be done when a task group chairperson is frequently absent from meetings?

A10. Sometimes this happens either because a chairperson is frequently away from the facility traveling or because she/he feels she/he is too busy. Regardless of the reason, top management should deal with the situation by encouraging the chairperson to become more active, or she/he should be replaced.

Q11. Can two task group members work on the same assignment?

A11. Most assignments can be managed by a single task group member. The primary reason for issuing assignments is to increase participation. As long as each task group member is actively participating in the management of the assignment, it should not present a problem if two task group members work on the same issue. But if an assignment is so large that additional help is needed, then it may be better to either break the assignment down into smaller, more easily managed components, or assign other employees to serve on a subcommittee chaired by the task group member who has been given the assignment. A good example would be an ad hoc committee to help organize and manage a poster contest.

Q12. How long will it take before we can expect to see results using the FSHC system?

A12. This will vary, but on average an employer should begin to see results in 4–6 months following implementation of the FSHC system. Following initial training to install the system, there is usually some confusion until task groups have the first meeting or two. By the time of the second meeting task groups and task group members have gained an understanding of their roles and what they should be doing. By the fourth or fifth meeting a change can usually be detected by a drop in injuries.

Q13. What will make this system fail?

A13. In a word, apathy. If top management is chairing the FSHC and makes his/her expectations clear to those involved, then the system will not fail. But if top management does not support the system, delegates leadership of the FSHC to an underling, or is perceived as less than committed to injury prevention, then the system will usually fail.

Q14. What is the most important feature of the system?

A14. Two issues are most critical: top management commitment and the assignments.

Q15. Should we rotate all task group members at the same time?

A15. It is better to have experienced task group members, so it is suggested that an employer establish a staggered rotation system and rotate half of the task group members each year.

Q16. Wouldn't it be better if all members of a task group worked together on the same assignment?

A16. This is old-school practice and would greatly reduce the effectiveness of the task groups. Employee safety–health doesn't wait for task groups to catch up. There is a constant need for inspections, training, and so on. Spreading out the assignments among 40 or 50 task group members allows the employer, through the FSHC system, to attack several fronts simultaneously and accomplish a great deal more, meeting the mandates of OSHA compliance but also protecting workers and improving awareness levels of task group members.

Q17. How many goals and/or objectives should a task group have?

A17. It is suggested that each task group establish from two to four goals and objectives at the beginning of each year. It is further suggested the facility EHS manager help each task group to determine, on the basis of top management's priorities, what those goals and objectives should be.

Q18. How are goals and objectives determined?

A18. Logic tells us that the best way to determine goals and objectives is to evaluate the needs of the company in terms of employee safety and health and environmental protection. A safety/health system audit might reveal poor inspections, poor training, a need for updated policies, and other deficiencies. While such an evaluation would be a good place to start, it is also important to meet with top management to determine what expectations are for EHS over the coming months.

Q19. What if we don't have enough managers to chair task groups?

A19. Obviously, among smaller employers this is a common issue. However, it poses little concern as everything is relative. If an employer has less than nine managers, then EHS issues are equally on a comparatively smaller scale. One way to handle this is to assign department managers to chair more than one task group, or employers can assign task group leadership to supervisors. However, this should not be done just for the sake of freeing a department manager of this responsibility.

Q20. Can task groups have co-chairs?

A20. This has been proved to be no less effective than single-chair task groups. Some employers, to satisfy labor contractual agreements, have appointed managers and employees to co-leadership roles for task groups. Where the common goal is incident prevention, this has been very effective.

Q21. Should employees other than task group members be involved?

A21. There are plenty of opportunities for those who are not currently serving on task groups. Employees and supervisors can serve on committees lead by task group members. They can lend their artistic and creative talents to contests, advertising, and other opportunities, for the Activities Task Group along with a host of other things. With an effective FSHC system, a safety culture is being created. By tapping the talents of non–task group members, an employer can get even more accomplished, build the culture faster, and at the same time prepare others for future roles as task group members.

Q22. Can I do the training and install the FSHC system myself?

A22. It certainly is possible to do this in-house; however, it is suggested that the training be conducted by someone who is experienced with the system and can anticipate problems that might arise, making recommendations to enable the employer to head these problems off before they become problems. While several employers have successfully installed the FSHC system described here, many more have tried it and failed.

Q23. How does the FSHC system differ from the behavior-based training that seems to be so popular?

A23. The behavior-based safety programs that so many employers are using have proved to be very effective in injury prevention at the outset; however, most of the employers who have used these programs as their sole safety–health program have encountered problems later on. Behavior based safety is not a safety program. It is not a safety system. It is a tool. It is something that must be used in conjunction with a managing system of some type in order to achieve long-term success. Those employers who have had long-term success using behavior-based safety (BBS) have effective managing systems in place and are not relying on BBS to be the entire safety–health program. Many employers have invested hundreds of thousands of dollars in BBS training only to have incidence rates climb back up after 3 or 4 years. The FSHC system is not the only way to manage EHS, but it is one of the best and most effective ways because it is a system and not a tool. Behavior-based training does no more than raise the awareness level of employees, just as this book has described. But that is where EHS management stops. To manage EHS successfully, an employer must incorporate EHS into the day-to-day management of the operations. That is what the system described in this book is intended to do.

Appendix C
Environmental, Safety, Health, and Security Resources

SERVICE ORGANIZATIONS

American National Red Cross National Headquarters
2025 E Street, NW
Washington, DC 20006
202-303-4498
URL: www.redcross.org

Prevent Blindness America
211 West Wacker Drive
Suite 1700
Chicago, IL 60606
URL: www.preventblindness.org

STANDARDS GROUPS

American National Standards Institute (ANSI)
25 West 43rd Street
New York, NY 10036
212-642-4900
URL: www.ansi.org

TRADE AND PROFESSIONAL ORGANIZATIONS

American Association of Occupational Health Nurses, Inc.
  (AAOHN)
2920 Brandywine Road, Suite 100
Atlanta, GA 30341
770-455-7757
URL: www.aaohn.org

American Board of Industrial Hygiene (ABIH)
6015 West St. Joseph, Suite 102
Lansing, MI 48917-3980
517-321-2638
URL: www.abih.org

American Chemical Society (ACS)
1155 16th Street, NW
Washington, DC
800-227-5558 (United States only)
202-872-4600 (outside the United States)
URL: www.acs.org

American College of Occupational and Environmental Medicine
  (ACOEM)
1114 North Arlington Heights Road
Arlington Heights, IL 60004-4770
847-818-1800
URL: www.acoem.org

American Conference of Governmental Industrial Hygienists
  (ACGIH)
1330 Kemper Meadow Drive
Cincinnati, OH 45240

Customers/members phone: 513-742-2020
Administrative phone: 513-742-6163
URL: www.acgih.org

American Industrial Hygiene Association (AIHA)
2700 Prosperity Avenue, Suite 250
Fairfax, VA 22031
703-849-8888
URL: www.aiha.org

American Medical Association (AMA)
515 North State Street
Chicago, IL 60610
800-621-8335
URL: www.ama-assn.org

American Petroleum Institute (API)
1220 L Street NW
Washington, DC 20005-4070
202-682-8000
URL: www.api.org

American Society of Mechanical Engineers (ASME)
3 Park Avenue
New York, NY 10016-5902
800-843-2763
URL: www.asme.org

American Society of Safety Engineers (ASSE)
1800 East Oakton
Des Plaines, IL 60018
847-699-2929
URL: www.asse.org

American Society for Testing and Materials (ASTM)
100 Barr Harbor Drive
West Conshohocken, PA 19428-2959
610-832-9555
URL: www.astm.org

American Welding Society (AWS)
550 NW LeJeune Road

Miami, FL 33126
800-443-9353
URL: www.aws.org

Board of Certified Hazard Control Management (BCHCM)
11900 Parklawn Drive, Suite 451
Rockville, MD 20852
301-770-2540
URL: www.chcm-chsp.org

Board of Certified Safety Professionals (BCSP)
208 Burwash Avenue
Savoy, IL 61874-9510
217-359-9263
URL: www.bcsp.org

Chemical Manufacturers Association, Inc. (CMA)
1300 Wilson Boulevard
Arlington, VA 22209
URL: www.cmahq.org
Compressed Gas Association (CGA)
4221 Walney Road, 5th floor
Chantilly, VA 20151-2923
703-788-2700
URL: www.cganet.com

Factory Mutual Engineering Organization (FM)
1151 Boston-Providence Turnpike
Norwood, MA 02062
781-762-4300
URL: www.factorymutual.com

Flight Safety Foundation, Inc.
601 Madison Street, Suite 300
Alexandria, VA 22314-1756
703-739-6700
URL: www.flightsafety.org

Human Factors and Ergonomics Society (HFES)
PO Box 1369
Santa Monica, CA 90406-1369
310-394-1811
URL: www.hfes.org

International Safety Equipment Association, Inc. (ISEA)
1901 North Moore Street, Suite 808
Arlington, VA 22209
703-525-1695
URL: www.safetyequipment.org

Material Handling Industry of America (MHIA)
8720 Red Oak Boulevard, Suite 201
Charlotte, NC 28217-3992
704-676-1190
URL: www.mhia.org

National Fire Protection Association (NFPA)
1 Batterymarch Park
Quincy, MA 02269-9101
617-770-3000
URL: www.nfpa.org

National Safety Council (NSC)
1121 Spring Lake Drive
Itasca, IL 60143-3201
630-285-1121
URL: www.nsc.org

National Safety Management Society (NSMS)
PO Box 4460
Walnut Creek, CA 94596-0460
800-321-2910
URL: www.nsms.us

Society of Automotive Engineers (SAE)
755 West Big Beaver, Suite 1600
Troy, MI 48084
248-273-2494
URL: www.sae.org

System Safety Society (SSS)
PO Box 70
Unionville, VA 22567-0070
540-854-8630
URL: www.system-safety.org

Underwriters Laboratories, Inc. (UL)
333 Pfingsten Road

Northbrook, IL 60062-2096
847-272-8800
URL: www.ul.com

Veterans of Safety (VOS)
CMSU, Missouri Safety Center
Humphreys #201
Warrensburg, MO 64093
660-543-4281

U.S. GOVERNMENT AGENCIES

Center for Disease Control (CDC)
http://www.cdc.gov

Code of Federal Regulations
http://www.access.gpo.gov/nara/cfr/index.html

Environmental Protection Agency (EPA)
http://www.eps.gov

National Highway Traffic Safety Administration (NHTSA)
http://www.nhtsa.dot.gov

National Institute of Occupational Safety and Health (NIOSH)
http://www.cdc.gov/niosh

Occupational Safety and Health Administration (OSHA)
http://www.osha.gov

U.S. Government Printing Office (GPO)
http://www.access.gpo.gov

CANADIAN SAFETY

Canada Safety Council (CSC)
1020 Thomas Spratt Place
Ottawa, Ontario, Canada K1G 5L5
613-730-1535
URL: www.safety-council.org

Canadian Standards Association (CSA)
5060 Spectrum Way
Mississauga, Ontario, Canada L4W 5N6
416-747-4000
800-463-6727
URL: www.csa.ca

Canadian Centre for Occupational Safety and Health (CCOHS)
135 Hunter Street East
Hamilton, Ontario, Canada L8N 1M5
1-800-668-4284
(toll-free in Canada and United States)
1-905-570-8094
http://www.ccohs.ca

Workplace Safety and Insurance Board (WS&IB)
200 Front Street West
Toronto, Ontario, Canada M5V 3J1
416-344-1000
Toll free: 1-800-387-5540
Ontario toll free: 1-800-387-0750
TTY: 1-800-387-0050
URL: www.wsib.on.ca

SECURITY INTERESTS

American Society for Industrial Security (ASIS)
1625 Prince Street
Alexandria, VA 22314-2818
703-519-6200
URL: www.asisonline.org

Association of Threat Assessment Professionals (ATAP)
PO Box 4108
Huntington Beach, CA 92605
310-312-0212
URL: www.atapusa.org

International Association of Professional Security Consultants (IAPSC)
525 SW 5th Street, Suite A

Des Moines, IA 50309-4501
515-282-8192
URL: www.iapsc.org

International Critical Incident Stress Foundation
3290 Pine Orchard Lane, Suite 106
Ellicott City, MD 21042
410-750-9600
URL: www.icisf.org

Index

Accident investigation, 108. *See also* Incident Investigation Task Group; Incidents

Accountability
employee, 1–2, 3, 4
importance of, 36
manager, 16

Activities
Fire and Emergency Task Group, 90–91
Health and Environment Task Group, 78–79
Housekeeping Task Group, 99–100
Incident Investigation Task Group, 108
with inspections, 72–73
safety-related, 39–40
Security Task Group, 116, 118

Annual emphasis programs, 45
Atmospheric testing, 12–13
Audit areas, 101
Audit procedures, establishing, 73
Audits, 72
electrical equipment, 73
fire prevention and protection, 96–97
housekeeping, 100–102
purposes of, 66–67
system, 73
Awards, housekeeping, 103
Awards/recognition programs, 46

Behavior-based safety (BBS), 199
Bottom-line management, 4–5
Boylston, Ray, ix–xi
Bulletin boards, 44

Effective Environmental, Health, and Safety Management Using the Team Approach, by Bill Taylor
Copyright © 2005 John Wiley & Sons, Inc.

Printed in the United States
By Bookmasters